# MAKING
# Algebra
# COME
# Alive

**Student Activities & Teacher Notes**

## ALFRED S. POSAMENTIER

**CORWIN PRESS, INC.**
A Sage Publications Company
Thousand Oaks, California

Copyright ©2000 by Corwin Press, Inc.

Material from *Principles and Standards for School Mathematics,* copyright 2000 by the National Council of Teachers of Mathematics, is reprinted by permission of the NCTM.

---

*For information:*

Corwin Press, Inc.
A Sage Publications Company
2455 Teller Road
Thousand Oaks, California 91320
E-mail: order@corwinpress.com

SAGE Publications Ltd
1 Oliver's Yard
55 City Road
London EC1Y 1SP

SAGE Publications In
B-42, Panchsheel Enc
Post Box 4109
New Delhi 110 017

Printed in the United States of America

**Library of Congress Cataloging-in-Publication Data**

Posamentier, Alfred S.
  Making algebra come alive: Student activities and teacher notes / by Alfred S. Posamentier.
      p. cm. — (Math assessment series)
   ISBN 0-7619-7596-9 (c) — ISBN 0-7619-7597-7 (p)
   1. Algebra—Problems, exercises, etc.  I. Title. II. Series.
   QA157 .P778 2000
   512—dc21                                         00-008374

This book is printed on acid-free paper.

00   01   02   03   04   05   10   9   8   7   6   5   4   3   2   1

---

| | |
|---|---|
| *Editorial Assistant:* | Catherine Kantor |
| *Production Editor:* | Diana E. Axelsen |
| *Editorial Assistant:* | Victoria Cheng |
| *Typesetter/Designer:* | Technical Typesetting, Inc. |
| *Designer:* | Tracy Miller |

# CONTENTS

# Introduction

*Making Algebra Come Alive* is a set of versatile enrichment exercises that cover a very broad range of mathematical topics and applications—from the Moebius strip to the googol. Several criteria have been used to develop the activities and to select the topics that are included. All of them bear heavily, and equally, on my concerns for curriculum goals and classroom management.

First and foremost, the activities are meant to be motivational. As much as possible, I want this book to achieve the goal of being attractive to people who thought they didn't like mathematics. To accomplish this, it is necessary for the activities to be quite different from what students encounter in their basal texts—different in both substance and form. This seems especially critical: No matter how excellently a basal text is being used, nearly every class experiences the "blahs." Unfortunately, this sort of boredom is often well entrenched long before the teacher and perhaps even the students are aware of it. Presenting activities on a regular basis gives the variety and change of pace needed to sustain interest in any subject.

With the number of topics you may have to cover during the normal school year, it may seem naïve or unrealistic to suggest introducing additional material. This brings me to the second criterion. Most of the activities in this volume can be used to enhance, reinforce, and extend the concepts and skills that already make up the better part of your curriculum and course goals. For some examples, see the activities in the problem-solving section of this book. These clearly reinforce your work in several areas. Similarly, the arithmetic operations section provides enhancement as well as variety to your work with computation skills. Thus, the activities are designed as an aid to presenting the basic concepts of your course, as well as a set of motivational and enrichment activities. These objectives are completely consistent with the *Principles and Standards for School Mathematics* (NCTM, 2000).

Third, it was felt that each activity should have some use or merit beyond itself, a heuristic value. That is, the activities serve as door openers; they can be introductions to areas not usually treated in basal texts. The activities provide good practice in what you're trying to teach anyway, but they also greatly increase your students' awareness of the different directions to which these ideas can lead.

## The Key: Problem Solving

Finally, the activities provide opportunities and incentives to hone problem-solving skills—not merely chapter-end exercises that are *often called* problems, but realistic problems such as your students will encounter in their everyday living and in later, nonmathematics school courses. Most of the activities begin by posing a problem that students find intriguing and that, at the outset, many students are unable to solve on their own. In working through the problem, however, the students discover they can tackle a much bigger monster than they thought they were capable of handling. Equally important, they find these problem-solving techniques are applicable to other areas.

The problem-solving orientation of these activities cannot be overemphasized. Those of you who are familiar with the NCTM Standards 2000 revision will find this book directly on target with most of the recommendations. Sometimes directly respondent, as with problem solving, and other times directly supportive, as with enrichment and applications, the *Math Alive Series* is a deliberate effort to meet the objectives of the Standards in a creative way.

## Presenting the Activities

In pilot testing these activities, I worked with teachers who had very diverse mathematical preparation and who had to deal with a *wide* spectrum of class size, student ability, and class heterogeneity. Thus, it seemed very desirable to search for alternative means to present the activities. I discovered several. One or more of them should be useful in your situation.

The normal presentation, the one that best suits most classes, is to present the activity as a new lesson at the outset of a class period. In working through the student page, you'll find the accompanying Teacher's Notes explain the rationale for the entire activity, as well as provide anticipated student responses and questions. However familiar you feel with the mathematical topics presented, do not attempt to conduct a class session without first spending 20 or 30 minutes going over the Teacher's Notes. Both the student pages and the Teacher's Notes are highly compressed: A typical student page encompasses the concepts that four or five basal-text pages generally treat.

In some cases, the student activity can be handed out the day preceding class discussion. Your perusal of the activity will best determine when this is appropriate. In many other cases, you will find it best to discuss only the body of the student activity the day you pass it out, deferring the discussion of the Extension until the following day.

If your class is like many that I have encountered, you may wish to try peer teaching. This has many advantages for both you and your students if your classes have three to six really bright students. By giving both the student page and Teacher's Notes to one of these "stars," he or she can present the activity the following day to this group of above-average students. This allows you time to work with your average and below-average students to bring their skills up to par without boring the students who are already well on top of things. My experience has shown that students who are asked to present activities prepare very well. Their pride is at stake, and thus you can be sure they won't let you down.

## The Extensions

The Extensions offer the greatest opportunity for flexibility in using the activities. Every activity in these volumes has one, but they differ. In some cases, they dip into more sophisticated mathematical concepts and should be considered as optional activities primarily for your better students. In other cases, the Extensions require no additional mathematical sophistication, but simply give an opportunity to explore the topic in greater detail. Your reading of the activity will quickly determine which is the case. Sometimes you may want to present the basic activity to the class and assign the Extension as homework for your better students. In all cases, you should think of the Extension as an element that allows you to tailor your mathematics program to best meet the needs and interests of all your students.

## Selecting the Activities by Topic

This volume probably contains more activities than you'll be able to use in a single school year and is probably sufficient to fully supplement the two-year sequence of algebra courses. The chapter introductions will assist you in selecting the activities best suited to your students' abilities and interests and offer some hints as to how they can be used. The activities have been divided into seven categories. These categories are not at all arbitrary, but your study of them will show that considerable overlap is possible. For example, "Many Happy Returns" in the Probability and Statistics section may be seen by some students as a recreational problem. For many of your students, the same will be true of the logic activities. Your better students will also consider all of the problem-solving activities to be recreational, but the problem-solving activities also contribute to students' abilities in all of the other areas.

In spite of these overlaps, the descriptions or categorizations can provide an aid to organizing your use of the activities. It is possible in many classes to use all of the activities by selecting the easier ones for your slower students and using the more difficult ones as extra-credit work for your better students.

## Selecting Activities by Level

Even the best two-year algebra sequences achieve only partial success in presenting topics in a logical order of prerequisites and difficulty. The historical growth of algebra and its subsequent absorption by contemporary cortex must follow a tree-like rather than a linear path. Because of this, some indication was given as to where in most courses these activities coincide with the concepts treated in most texts. However, I haven't strongly emphasized this consideration. Instead, I have stated the principles and operations students need to successfully tackle an activity. This of course shifts the emphasis to allow the activities to be used as change-of-pace endeavors that do not impede your progress toward completing the syllabus. The probability and logic activities are prime examples of this. These topics are usually presented late in the textbook sequence, but they can be treated much earlier.

The following pages give an overview of the activities, category by category. A star (⋆) indicates the activities that are probably best given only to your better students or given a more careful presentation to the general class. The other volumes in this series use a diamond (♦) to indicate activities more suitable for students who are finding the sledding a little rough. Here, however, it is enough to refer you to *Making Pre-Algebra Come Alive*.

## The NCTM Principles and Standards for School Mathematics – 2000

Each unit is tied in with one or more of the NCTM Standards presented in *Principles and Standards for School Mathematics – 2000*. As units are selected for use in the classroom, it is good to be aware of the Standards being employed. A simple numbering system is used to help make this identification simple and unobtrusive. At the start of each "Teacher Notes" section, the Standard number appropriate for that unit is indicated by a dot below the appropriate Standard number. These numbers correspond to the following list of standards:

### 1. Number and Operations Standard

Instructional programs from prekindergarten through grade 12 should enable all students to –

- understand numbers, ways of representing numbers, relationships among numbers, and number systems;
- understand meanings of operations and how they relate to one another;
- compute fluently and make reasonable estimates.

### 2. Algebra Standard

Instructional programs from prekindergarten through grade 12 should enable all students to –

- understand patterns, relations, and functions;
- represent and analyze mathematical situations and structures using algebraic symbols;
- use mathematical models to represent and understand quantitative relationships;
- analyze change in various contexts.

### 3. Geometry Standard

Instructional programs from prekindergarten through grade 12 should enable all students to –

- analyze characteristics and properties of two- and three-dimensional geometric shapes and develop mathematical arguments about geometric relationships;

- specify location and describe spatial relationships using coordinate geometry and other representational systems;
- apply transformations and use symmetry to analyze mathematical situations;
- use visualization, spatial reasoning, and geometric modeling to solve problems.

## 4. Measurement Standard

Instructional programs from prekindergarten through grade 12 should enable all students to –

- understand measurable attributes of objects and the units, systems, and processes of measurement;
- apply appropriate techniques, tools, and formulas to determine measurements.

## 5. Data Analysis and Probability Standard

Instructional programs from prekindergarten through grade 12 should enable all students to –

- formulate questions that can be addressed with data and collect, organize, and display relevant data to answer them;
- select and use appropriate statistical methods to analyze data;
- develop and evaluate inferences and predictions that are based on data;
- understand and apply basic concepts of probability.

## 6. Problem Solving Standard

Instructional programs from prekindergarten through grade 12 should enable all students to –

- build new mathematical knowledge through problem solving;
- solve problems that arise in mathematics and in other contexts;
- apply and adapt a wide variety of appropriate strategies to solve problems;
- monitor and reflect on the process of mathematical problem solving.

## 7. Reasoning and Proof Standard

Instructional programs from prekindergarten through grade 12 should enable all students to –

- recognize reasoning and proof as fundamental aspects of mathematics;
- make and investigate mathematical conjectures;
- develop and evaluate mathematical arguments and proofs;
- select and use various types of reasoning and methods of proof.

## 8. Communication Standard

Instructional programs from prekindergarten through grade 12 should enable all students to –

- organize and consolidate their mathematical thinking through communication;
- communicate their mathematical thinking coherently and clearly to peers, teachers, and others;
- analyze and evaluate the mathematical thinking and strategies of others;
- use the language of mathematics to express mathematical ideas precisely.

## 9. Connections Standard

Instructional programs from prekindergarten through grade 12 should enable all students to –

- recognize and use connections among different mathematical ideas;
- understand how mathematical ideas interconnect and build one another to produce a coherent whole;
- recognize and apply mathematics in contexts outside of mathematics.

## 10. Representation Standard

Instructional programs from prekindergarten through grade 12 should enable all students to –

- create and use representations to organize, record, and communicate mathematical ideas;
- select, apply, and translate among mathematical representations to solve problems;
- use representations to model and interpret physical, social, and mathematical phenomena.

# ABOUT THE AUTHOR

Alfred S. Posamentier is Professor of Mathematics Education and Dean of the School of Education of the City College of the City University of New York. He is the author and coauthor of numerous mathematics books for teachers and secondary school students. As a guest lecturer, he favors topics regarding aspects of mathematics problem solving and the introduction of uncommon topics into the secondary school realm for the purpose of enriching the mathematics experience of those students. The development of this book reflects these penchants.

After completing his A.B. degree in mathematics at Hunter College of the City University of New York, he took a position as a teacher of mathematics at Theodore Roosevelt High School in the Bronx (New York), where he focused his attention on improving the students' problem-solving skills. He also developed the school's first mathematics teams (at both the junior and senior level) and established a special class whose primary focus was on mathematics problem solving and enrichment topics in mathematics.

For years, Dr. Posamentier has collected clever ways of introducing students to new concepts in mathematics. This collection of ideas prompted the development of this book. He is currently involved in working with mathematics teachers, locally and internationally, to help them better understand problem-solving strategies and alternative instructional strategies, so that they can comfortably incorporate them into their regular instructional program.

Immediately upon joining the faculty of the City College (after having received his masters' degree there), he began to develop inservice courses for secondary school mathematics teachers, including such special areas as recreational mathematics, problem solving in mathematics, and instructional alternatives for the classroom.

Dr. Posamentier received his Ph.D. from Fordham University (New York) in mathematics education and has since extended his reputation in mathematics education to Europe. He is an Honorary Fellow at the South Bank University (London, England). He has been visiting professor at several Austrian, British, and German universities, most recently at the University of Vienna and at the Technical University of Vienna. At the former, he was a Fulbright Professor in 1990.

In recognition of his outstanding teaching, the City College Alumni Association named him Educator of the Year and had a day (May 1, 1993) named in his honor by the City Council President of New York City. In 1994, he was awarded the National Medal of Honor from the Austrian government, and in 1999, by an act of the Austrian Parliament he was awarded an Austrian University Professor title by the President of the Republic of Austria.

Naturally, with his penchant for motivating students towards mathematics, he has been very concerned that students have a proper introduction to mathematics from an entertaining point of view. This interest motivated the development of this book.

# CHAPTER 1

# Number Theory

- Number Classifications
- Getting on Base
- Friendly Numbers
- The Fibonacci Sequence ★
- Pascal's Triangle ★

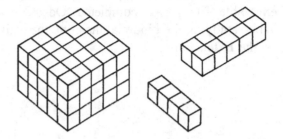

The activities in this section present "number theory" topics in an algebraic context. Thus, the problems here allow students to become acquainted with elementary algebraic operations while dealing with some patterns and terms that they've encountered before. The first two activities are particularly good to present during the first four weeks of an Algebra I course.

"Number Classifications" is the mathematical equivalent of a very brief course in taxonomy. Rather than simply giving examples and asking for rote memorization, this activity has students derive the examples by solving simple equations. The use of a number line and Venn diagram adds further "tangibility" to the taxonomy of numbers. The Extensions give additional flexibility: Depending on where in your course this activity is used, you can discuss imaginary numbers to whatever extent is appropriate. "Number Classifications" is also an excellent choice for the beginning of an Algebra II course because most texts begin by reviewing number classifications.

"Getting on Base" takes a sometimes dull and often confusing topic and presents it in an operationally clear manner. If a few additional examples are given, most students attain proficiency in base conversions very quickly. More importantly, the exercises in the latter half of this activity are riddles that students really enjoy. This is a very popular activity for use early in the algebra course.

The title "Friendly Numbers" is somewhat misleading. It connotes a relatively easy activity, and this really isn't the case here. Although an ability to find all the factors of a number is the only prerequisite for the activity, students must enjoy arithmetic manipulations to be turned on by this one. For this predisposed group, however, the activity is a lot of fun.

As previously mentioned, most students study mathematics to solve problems in areas that are not purely mathematical. Thus, your budding biology students will be pleased to find a clear biological application in the work of the thirteenth-century mathematician Leonardo of Pisa, known as Fibonacci. "The Fibonacci Sequence" also emphasizes deriving generalizations from specific patterns of sums of terms. The mathematical notation and the proofs given in the Teacher's Notes will help prepare your better students for more advanced mathematics.

Because it combines aesthetic simplicity with enormous versatility, "Pascal's Triangle" is a favorite of nearly everyone. If the Extension is omitted, "Pascal's Triangle" can be used anywhere in the Algebra I course. However, one of the most mathematically delightful aspects of this activity is the range of topics related to Pascal's triangle. Thus, students will probably find the activity most enjoyable if they have completed "Odds Are..." and "The Fibonacci Sequence." When studying the binomial theorem, the activity can be repeated with emphasis on the Extension.

# Number Classifications

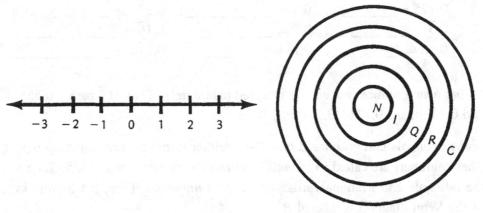

In the equation $3x + 5 = 11$, $x =$ _____.

What kind of number is this? _____

Let's change the equation a little so that it is $3x + 11 = 5$.

Now $x =$ _____. What kind of number is this?

_____

What is the value of $x$ in $2x + 1 = 6$? _____

What kind of number is this? _____

So far you've uncovered three different kinds of numbers. Your first value for $x$ is a *natural number*. These are the numbers $1, 2, 3, \ldots$ and are the numbers we use for counting.

How many members are in this set? _____

Your second value of $x$ is an *integer*. The integers are on _____ sides of the number line, whereas natural numbers lie only on the _____ side of the number line. Thus, natural numbers are a *subset* of the set we call integers, and so in the Venn diagram on this page, the circle labeled $I$ includes the circle labeled $N$.

Finally, in finding $x = 5/2$, you have come up with a number that is neither a natural number nor an integer. These are numbers formed by the ratio of two integers, $a/b$, where $b \neq 0$. Because they are derived from ratios, they are called *rational numbers* and are represented by $Q$ (for quotient) on the Venn diagram. How is it that circle $Q$ can enclose circles $N$ and $I$? _____

_____

How many rational numbers exist on the number line between (say) points 2 and 3? _____

This takes care of fractions. What about decimals? Most decimals can be converted to fractions that are the exact equivalent, so decimals are rational numbers too. Try the following:

$0.125 =$ _____ ; $0.6 =$ _____ ; $0.375 =$ _____ ;
$0.\overline{3} =$ _____ ; $0.\overline{5} =$ _____ ; $0.1\overline{6} =$ _____ ;
$0.8\overline{3} =$ _____ ; $0.\overline{13} =$ _____ ; $0.\overline{27} =$ _____ .

(A bar over a number means it is repeated indefinitely. Thus, $0.\overline{5}$ means 0.555..., and $0.\overline{13}$ means 0.131313....)

Some decimals can go on and on. They neither terminate nor infinitely repeat. These numbers are called *irrationals*. Examples of these are $\pi$ and $\sqrt{5}$. Together, the rationals and irrationals make up the *real numbers*. Thus, the fourth circle of the Venn diagram is labeled $R$.

**EXTENSION!** What is the solution to the equation $x^2 + 7 = 3$? What kind of number is this?

# Teacher's Notes for Number Classifications

*Many students have trouble recalling the distinctions among natural, integral, rational, and real numbers. This is to be expected because the time between encountering these terms is usually large compared to the frequency with which the students use the terms. As a result, whenever discussions of generalizations or abstractions using these terms are presented, they may be a small stumbling block, and these terms have to be defined. This activity shows the reasoning by which these terms have been named and ordered. The use of a number line and a Venn diagram also helps students keep the terms straight. Ask students to show their work.*

--- NCTM Standards ---

| 1 | 2 | 3 | 4 | 5 | 6 | 7 | 8 | 9 | 10 |
|---|---|---|---|---|---|---|---|---|----|
| • | • |   |   |   | • | • | • |   |    |

### Presenting the Activity

When students solve the first equation they find $x = 2$. They may answer "positive number" or "whole number" to describe what kind of number this is. However, because these numbers are our "counting numbers"—the first kind of number that humans thought about—we call them *natural numbers*. This set of numbers, $N = 1, 2, 3, \ldots$, is infinite, ordered, and has a first term, 1.

The second equation yields $x = -2$. This number belongs to the set of *integers*

$$I = \ldots, -3, -2, -1, 0, 1, 2, 3, \ldots$$

and includes the left as well as the right side of the number line. This set is also infinite and ordered, but it has no first or smallest term.

The solution to the third equation is a fraction, 5/2. This belongs to the set of *rational numbers*, that, because it includes fractions, is larger than the set of integers. A few students may wonder why this set includes the set of integers. Point out that because these numbers are defined as $a/b$ ($b \neq 0$), an integer can result from (say) 6/3. Of course, there's no law saying $a$ can't equal $b$, yielding 1. Discuss, too, that these terms are "everywhere dense" along the number line; an infinite number of rational numbers exist between any two specific values on the number line. As an example, between 3/32 and 1/8 you can indefinitely divide the interval by 2, getting 7/64, and so on.

Most students quickly recognize that numbers such as 0.125 and 0.375 are easily converted to 1/8 and 3/8 and thus fit the classification of rational numbers. They'll be a little less certain about numbers such as $0.1\overline{6}$ and $0.\overline{27}$. However, converting these decimals to fractions is quite easy, as shown by

$$
\begin{array}{ll}
N = 0.555\ldots & \\
10N = 5.555\ldots & \times\ 10 \\
\underline{-N = 0.555\ldots} & \text{subtract} \\
9N = 5.000 & \div\ 9 \\
N = \frac{5}{9} &
\end{array}
\qquad
\begin{array}{l}
N = 0.1666\ldots \\
10N = 1.6666\ldots \\
\underline{-N = 0.1666\ldots} \\
9N = 1.5000 \\
N = \frac{1.5}{9} = \frac{1}{6}
\end{array}
$$

5

$$
\begin{array}{llcl}
N = & 0.1313\ldots & & N = 0.2727\ldots \\
100N = & 13.1313\ldots & \times\ 100 & 100N = 27.2727\ldots \\
-\ \ N = & 0.1313\ldots & \text{subtract} & -\ \ N = 0.2727\ldots \\
\hline
99N = & 13 & \div\ 99 & 99N = 27 \\
N = & \frac{13}{99} & & N = \frac{27}{99} = \frac{3}{11}
\end{array}
$$

Thus, all terminating and repeating decimals represent rational numbers.

Nonterminating, nonrepeating decimals, however, are *irrational*. Point out that 22/7, 3.14, and 3.1416 are simply *rational approximations* of the irrational number $\pi$. The same is true for $\sqrt{5} = 2.236$. These numbers are *real*—unlike imaginary numbers, they do exist on the number line. We just can't pinpoint *exactly* where (although we can come very close). Have students look at a table of square roots. They will discover that only the square roots that contain a repetend in their decimal representations are those of perfect squares. All other square roots are irrational numbers, because they are nonterminating, nonrepeating decimals.

### Extension

You can round out your activity of number classifications with a very brief discussion of *complex numbers*. Explain to students that $\pm 2$ cannot be the solution to the equation; no real number multiplied by itself equals $-4$. Thus, the symbol $i = \sqrt{-1}$ was devised to represent these *imaginary numbers*. These do not exist on the number line. Together with real numbers, imaginary numbers form the set of complex numbers. These are not simply mathematical oddities whose only function is to smooth over a quirk in the number system; they're indispensable for electromagnetic systems calculations.

If you wish, you can have students make up mnemonic sentences to keep the order of *NIQRC* straight. They can be silly, of course, such as "Nine Idiots Quickly Reached College." The only requirement is that the sentences be easily remembered.

### Alternate Extension

You may want to use a quiz such as the following instead of or in addition to the Extension. You can use your best judgment as to whether to use the complex examples.

1. Identify the following numbers as belonging to the set of natural numbers, integers, rational numbers, real numbers, or complex numbers (name the "smallest" possible set in each case): $-3, 5/3, 17, \sqrt{2}, 3.14, 22/7, \sqrt{-9}, 0.\overline{4}, 0.21333\ldots, 0.71828\ldots,$ $0.\overline{15}, 0.1\overline{5}, -1/4, -\sqrt{16}$, etc.

2. Convert each of the following fractions to decimals: $\frac{3}{8}, \frac{7}{5}, \frac{2}{3}, \frac{7}{9}, \frac{5}{11}, \frac{5}{12}$.

3. Convert each of the following decimals to fractions: $0.875, 0.\overline{8}, 0.363636\ldots,$ $0.83333\ldots$.

# Getting on Base

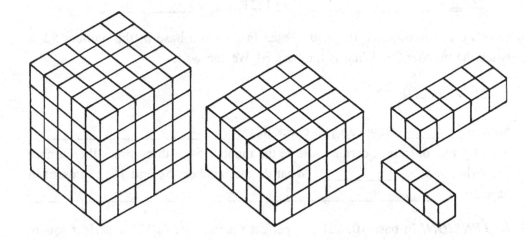

You know that a number such as 356 doesn't stand for $3 + 5 + 6$. It stands for $3 \times$ _____ $+ 5 \times$ _____ $+ 6 \times$ _____. Similarly, 2674 represents $2 \times$ _____ $+ 6 \times$ _____ $+ 7 \times$ _____ $+ 4 \times$ _____. In each case we multiply by successive powers of the base. Our system is based on 10. So, moving from right to left, we multiply by $10^0$, $10^1$, $10^2$, $10^3$, and so forth.

We do the same thing no matter what base is being used. If humans had evolved with eight fingers and toes, we might have a base-8 system. In this system, our 2674 (now written $2674_{\text{eight}}$) would mean

$$2 \times 8^3 + 6 \times \text{_____} + 7 \times \text{_____} + 4 \times \text{_____}.$$

This is $1024 + 384 + 56 + 4 = 1468$ in our base-10 system. We can say then that $1468_{\text{ten}} = 2678_{\text{eight}}$. What is $356_{\text{eight}}$ in our base-10 system? _____

Now let's try going the other way. How can we express 816 in base 5? Begin by finding the following powers of 5:

$5^0 =$ _____, $5^1 =$ _____, $5^2 =$ _____,

$5^3 =$ _____, $5^4 =$ _____, $5^5 =$ _____.

Whereas 816 is between $5^4$ and $5^5$, our first digit will be a multiple of $5^4$. How many 625s are contained in 816? _____ Therefore, our first digit will be _____. How many digits will the number have? _____ Now subtracting 625 from 816 gives _____. How many $5^3$ are contained in this? _____ Continue in this way until you find $816_{\text{ten}} =$ _____five.

Now try these:

$262_{ten} = $ _____ four, $\quad$ $12212_{three} = $ _____ ten,

$262_{ten} = $ _____ six, $\quad\quad$ $12212_{five} = $ _____ ten,

$262_{ten} = $ _____ seven, $\quad$ $12212_{eight} = $ _____ ten.

Now we can have some fun with these. In a certain base $b$, the number 52 is twice the number 25. What is the base $b$? We can write

$$5b + 2 = 2(2b + 5).$$

So $b = $ _____. How can you check this? _____

Now try this one: In a certain base $b$, the number 54 is three times the number 16. What is $b$? _____ What do the numbers $5_b$ and $16_b$ represent in base 10? _____

*EXTENSION!* In base 10, 121 is a perfect square, $11^2$. Is 121 a perfect square in any other base? What about 144 and 169 ($12^2$ and $13^2$)?

# Teacher's Notes for Getting on Base

*A good case can be made for students having some awareness that other bases are possible. (A small, but active society is still promoting worldwide adoption of base 12.) In addition, while studying facets of other bases, students do indeed acquire a better sense of place value in base 10. The problems at the end of this activity and its Extension provide some pleasant, eye-opening surprises for your better students.*

| | | | | | NCTM Standards | | | | | |
|---|---|---|---|---|---|---|---|---|---|---|
| 1 | 2 | 3 | 4 | 5 | 6 | 7 | 8 | 9 | 10 |
| • | • | • | | | | | | | |

### Presenting the Activity

The activity begins with straightforward exercises in converting numbers from base 10 to other bases and vice versa. How much time you allow for this part of the activity will depend on how much exposure your students have had to conversions between number bases and how recent their practice has been.

Before trying to state what a base-10 number would represent if we assume its digits reflect base $b$, students should check to be certain that the largest digit is at least 1 smaller than the base itself. Thus, our 2674 could represent a base-8 number, but not any base smaller than that.

In finding what a base-10 number represents if its digits are thought of as base-8 numbers, we simply work from right to left, multiplying the digits by successively higher powers of the new base. Thus, $2674_{\text{eight}}$ is

$$4 \times 8^0 + 7 \times 8^1 + 6 \times 8^2 + 2 \times 8^3 = 1468_{\text{ten}}$$

and $356_{\text{eight}}$ is

$$6 \times 8^0 + 5 \times 8^1 + 3 \times 8^2 = 238_{\text{ten}}.$$

To work the other way, begin by expanding whatever base you're using until you reach a number larger than the number you want to convert. For example, $5^0 = 1$, $5^1 = 5$, $5^2 = 25$, $5^3 = 125$, $5^4 = 625$, $5^5 = 3125$. Because 3125 is larger than 816, the first digit will be a multiple of $5^4$ and the number will have five digits, some multiples of $b^0 + b^1 + b^2 + b^3 + b^4$. Subtract 625 from 816 and repeat the process with the difference. Our result then is:

$$1(625) + 1(125) + 2(25) + 3(5) + 1 = 11231_{\text{five}}.$$

For the remaining examples on the student page,

$$262_{\text{ten}} = 1(4)^4 + 0(4)^3 + 0(4)^2 + 1(4)^1 + 2(4)^0 = 10012_{\text{four}},$$
$$262_{\text{ten}} = 1(6)^3 + 1(6)^2 + 1(6)^1 + 4(6)^0 = 1114_{\text{six}},$$
$$262_{\text{ten}} = 5(7)^2 + 2(7)^1 + 3(7)^0 = 523_{\text{seven}},$$
$$12212_{\text{three}} = 1(3)^4 + 2(3)^3 + 2(3)^2 + 1(3)^1 + 2(3)^0 = 158_{\text{ten}},$$

$$12212_{\text{five}} = 1(5)^4 + 2(5)^3 + 2(5)^2 + 1(5)^1 + 2(5)^0 = 932_{\text{ten}},$$
$$12212_{\text{eight}} = 1(8)^4 + 2(8)^3 + 2(8)^2 + 1(8)^1 + 2(8)^0 = 5258_{\text{ten}}.$$

The last two problems on the student page have a riddle-like quality that many people enjoy. You may wish to present several others to the class. For the first problem, $5b + 2 = 2(2b + 5)$. Thus $b = 8$. This can easily be checked by substituting the base into the equation, which, in base 10, gives $42 = 42$.

The second problem is similar:

$$5b + 4 = 3(b + 6).$$

This yields $b = 7$. Checking as before, $39 = 39$.

Finally, some of your students may be interested in seeing some special conversions wherein one base is an integral power of the other base. Thus,

$$101110100111_{\text{two}} = \underbrace{10}_{2}\ \underbrace{11}_{3}\ \underbrace{10}_{2}\ \underbrace{10}_{2}\ \underbrace{01}_{1}\ \underbrace{11}_{3}$$
$$= 232213_{\text{four}}$$

$$101110100111_{\text{two}} = \underbrace{101}_{5}\ \underbrace{110}_{6}\ \underbrace{100}_{4}\ \underbrace{111}_{7}$$
$$= 5647_{\text{eight}}$$

### Extension

Investigate this problem as follows:

$$121_b = 1(b)^2 + 2(b)^1 + 1(b)^0$$
$$= b^2 + 2b + 1$$
$$= (b + 1)^2.$$

Surprise! The numeral 121 represents a perfect square in *any* positive base $b \geqslant 3$, and is the square of one more than the base number.

Students may discover other such numbers by squaring $b + 2$ and $b + 3$ to obtain 144 and 169. These perfect squares in base 10 are also perfect squares in any positive integral base that contains the digits used in them ($b \geqslant 5$ and $b \geqslant 10$, respectively). It's not necessary for the coefficient of $b$ to equal 1. If you square $2b + 1$, for example, you obtain $4b^2 + 4b + 1 = 441_b$, which will be a perfect square in any positive integral base $b \geqslant 5$. Have students square other expressions, such as $3b + 1$, $2b + 2$, and $4b + 1$, to obtain other perfect squares.

Some students may wish to search for perfect cubes, perfect fourth powers, and so on. By cubing $b + 1$, for example, they obtain $b^3 + 3b^2 + 3b + 1$. This indicates that 1331 is a perfect cube in any positive integral base $b \geqslant 4$. (In base 10, $1331 = 11^3$.) In each case, 1331 is the cube of one more than the base number.

# Friendly Numbers

If Lisa wore a button with the number 220 on it and David wore a similar button with the number 284, we might wonder what that was supposed to represent. It turns out that 220 and 284 are examples of *friendly numbers*, so we can assume that Lisa and David like each other. Let's see why these two numbers are considered "friendly."

List all the factors of 220 (less than 220) and find their sum.

Factors: ____ ____ ____ ____ ____ ____ ____ ____ ____ ____

Sum: _____

Now find the sum of the factors of 284 (less than 284).

Factors: ____ ____ ____ ____ ____

Sum: _____

What unusual relationship appears to exist here? _____

_____

_____

It isn't easy to find friendly numbers. The famous Swiss mathematician Leonhard Euler (1707–1783) listed 60 pairs of friendly numbers in 1750. Strangely enough, he omitted the second smallest pair. This pair was later discovered by a 16-year-old boy, Nicolo Paganini, in 1866. If one number of this friendly pair of numbers is 1184, what is its mate? _____

How can you check whether this number is in fact the other member of a friendly pair? _____

_____

_____

Suppose we now try to determine whether 6 is a member of a friendly pair. The sum of the factors of 6 (less than 6) is _____ .

Any number that is "friendly to itself" is called a *perfect number*.

Find the factors of 28 (less than 28). Is 28 a perfect number? _____

Find the friendly mate of 496. How would you classify this number? _____

***EXTENSION!*** Here is a formula for generating even perfect numbers $P$:

$$P = (2^{n-1})(2^n - 1).$$

Use this formula to find other perfect numbers. Use these values for $n$: 2, 3, 5, 7, 13, and 17.

# Teacher's Notes for Friendly Numbers

*Patterns and relationships between numbers are often fascinating and surprising, as this activity proves. Students can become so involved with operations on numbers, they never discover the interesting properties of numbers.*

*The only prerequisite for this activity is finding all the factors of a number.*

_____ NCTM Standards _____

| 1 | 2 | 3 | 4 | 5 | 6 | 7 | 8 | 9 | 10 |
|---|---|---|---|---|---|---|---|---|----|
| • | • |   |   |   |   |   |   |   |    |

### Presenting the Activity

Before presenting the activity, review how to find all the factors of a number if students have not studied it recently.

Be sure students understand that they are to find all the factors of the number *except* the number itself. The factors of 220 (less than 220) are 1, 2, 4, 5, 10, 11, 20, 22, 44, 55, and 110. The sum of the factors is 284. The factors of 284 (less than 284) are 1, 2, 4, 71, and 142. Their sum is 220. Thus, the sum of the factors of one friendly number (not including the number itself) is the other friendly number of the pair.

The factors of 1184 (less than 1184) are 1, 2, 4, 8, 16, 32, 37, 74, 148, 296, and 592. Their sum is 1210. Thus, 1184 and 1210 are friendly numbers if the reverse is also true. To verify this, students should find the sum of the factors of 1210 (less than 1210). The factors are 1, 2, 5, 10, 11, 22, 55, 110, 121, 242, and 605. Their sum is indeed 1184. Indeed they *are* friendly numbers.

Students may be interested in other pairs of friendly numbers. By 1972, 1107 friendly number pairs were known. The largest of these is

    4,522,265,534,545,208,537,974,785    and
    4,539,801,326,233,928,286,140,415.

Others are 2620 and 2924, 5020 and 5564, and 6232 and 6368. There is no single formula for finding friendly pairs of numbers.

The sum of the factors of 6 (less than 6) is

    $1 + 2 + 3 = 6.$

Thus, 6 is a perfect number. The factors of 28 (less than 28) are 1, 2, 4, 7, and 14. Their sum is 28, so 28 is also a perfect number. The factors of 496 (less than 496) are 1, 2, 4, 8, 16, 31, 62, 124, and 248. Their sum is 496, so 496 is also a perfect number.

### Extension

The perfect numbers generated by the formula and the values of *n* given on the student page are

| $n$ | $P$ |
|---|---|
| 2 | 6 |
| 3 | 28 |
| 5 | 496 |
| 7 | 8128 |
| 13 | 33,550,336 |
| 17 | 8,589,869,056 |

These values of $n$ generate the first six perfect numbers. For the formula to work, $n$ must be a prime number and $(2^n - 1)$ must be a prime number.

The next two perfect numbers are 137,438,691,328 and 2,305,843,008,139,952,128. They are generated using 19 and 31 for $n$.

You may wish to point out that no *odd* perfect numbers are known. Also, there is no more than one perfect number between any two consecutive powers of 10.

As noted previously, the factor $(2^n - 1)$ must be a prime. Numbers in the form $2^n - 1$, where $n$ is prime, are called *Mersenne* numbers. The first eight Mersenne numbers are

| $n$ | $M = (2^n - 1)$ |
|---|---|
| 2 | 3 |
| 3 | 7 |
| 5 | 31 |
| 7 | 127 |
| 13 | 8191 |
| 17 | 131,017 |
| 19 | 524,287 |
| 31 | 2,147,483,647 |

# The Fibonacci Sequence

Everyone knows that rabbits can reproduce at a very fast rate. You may be surprised to learn that an Italian mathematician, Leonardo of Pisa, also known as Fibonacci, was interested in the rate of rabbit reproduction. In 1202 he published a book that contained the following problem:

> How many pairs of rabbits will be produced in a year, beginning with a single pair, if in one month each pair bears a new pair which becomes productive from the second month on?

| Month | | No. of Pairs |
|-------|---|---|
| Jan. | B | 1 |
| Feb. | A | 1 |
| Mar. | A B | 2 |
| Apr. | A B A | 3 |
| May | A B A A B | 5 |
| June | A B A A B A B A | 8 |
| July | | |

The accompanying chart shows the pattern of rabbit reproduction for the first six months of the year. *B* stands for a *pair* of baby rabbits that cannot reproduce yet and *A* stands for a *pair* of adult rabbits that will produce a pair of baby rabbits the next month. Fill in the chart for July. How many pairs of rabbits will there be in August? _____ How many in September? _____

The sequence of numbers in the right-hand column of the chart is called the *Fibonacci sequence*. Fill in the missing values in the sequence

$$1, 1, 2, 3, 5, 8, \underline{\quad}, \underline{\quad}, \underline{\quad}, \ldots .$$

How is each term of the sequence found? _____

Write the first 12 terms of the Fibonacci sequence: _____

How many pairs of rabbits will there be in a year? _____

If $f_7$ represents the seventh term of the Fibonacci sequence, then $f_7 = f_5 + f_6$. Using this notation, how would you write the $n$th term, $f_n$, of the sequence?

_____

Now find the sum of the first four terms of the Fibonacci sequence:

$$f_1 + f_2 + f_3 + f_4 = 1 + 1 + 2 + 3 = \underline{\hspace{1.5cm}}.$$

How does this number compare to $f_6$, the sixth term? _____

For $n = 7$, find $f_1 + f_2 + \cdots + f_n = \underline{\hspace{1.5cm}}$. What is $f_{n+2}$? _____

How would you write the sum of the first $n$ terms, $f_1 + f_2 + \cdots + f_n$? _____

There are other interesting sums in the Fibonacci sequence. Let's consider every other number beginning with the first number: $f_1 + f_3 + f_5 + f_7 + f_9$.

These are the first five numbers with odd *indices*. What is their sum? _____

Which term in the Fibonacci sequence is the same as this sum? _____

Find this sum without adding: $f_1 + f_3 + f_5 + f_7 + f_9 + f_{11} = \underline{\hspace{2.5cm}}.$

Now consider every other number beginning with the second number. These are the numbers with even indices. What is the sum of the first four Fibonacci numbers with even indices? _____ How does this number compare to $f_9$? _____ Find this sum without adding: $f_2 + f_4 + f_6 + f_8 + f_{10} = \underline{\hspace{2.5cm}}.$

***EXTENSION!*** Experiment with specific cases and find the sum $f_1^2 + f_2^2 + f_3^2 + \cdots + f_n^2$ in terms of the product of two Fibonacci numbers.

# Teacher's Notes for the Fibonacci Sequence

*This activity introduces the Fibonacci sequence and explores some of its properties. Some students may be familiar with sequences from their work in earlier courses, but will be surprised to find a sequence that reflects a pattern in nature. Other Fibonacci numbers in nature are explored in "Mathematics in Nature" in* **Making Pre-Algebra Come Alive.**

*This activity emphasizes using mathematical notation to discover the properties of the Fibonacci sequence and will help students make generalizations in other situations.*

--- NCTM Standards ---

| 1 | 2 | 3 | 4 | 5 | 6 | 7 | 8 | 9 | 10 |
|---|---|---|---|---|---|---|---|---|----|
| • | • | • |   |   |   |   |   | • | •  |

### Presenting the Activity

After students have read through the problem posed by Fibonacci, go over the chart in detail to be sure they understand it. Point out that after a pair of rabbits are adult, the adult pair and the pair of babies they produce will be counted the next month. It may be necessary to draw the figure on the chalkboard and complete the row for July that shows 13 pairs of rabbits. In August there will be 21 pairs and in September, 34 pairs.

The sequence that represents the numbers found through September is 1, 1, 2, 3, 5, 8, 13, 21, 34, .... Most students should see on their own that each term of the sequence after the second term is the sum of the two preceding terms. Be sure all students realize this before proceeding with the activity. Students can now easily write the first 12 terms of the Fibonacci sequence—1, 1, 2, 3, 5, 8, 13, 21, 34, 55, 89, 144—and see that there will be 144 pairs of rabbits in a year.

Be sure students understand the notation used to represent the terms of the sequence. The $n$th term $f_n = f_{n-1} + f_{n-2}$. The sum of the first four terms is 7, one less than the sixth term. Similarly, $f_1 + f_2 + \cdots + f_7 = 33 = f_9 - 1$ and, in general form, $f_1 + f_2 + \cdots + f_n = f_{n+2} - 1$. This can be shown algebraically as

$$f_1 = f_3 - f_2 \quad \text{(since } f_3 = f_1 + f_2 \text{)}$$
$$f_2 = f_4 - f_3$$
$$f_3 = f_5 - f_4$$
$$\vdots$$
$$f_{n-1} = f_{n+1} - f_n$$
$$f_n = f_{n+2} - f_{n+1}.$$

If these equations are added, the sum on the left side is $f_1 + f_2 + f_3 + \cdots + f_n$. On the right side, all the terms except two cancel out and we have $f_{n+2} - f_2$. We know that $f_2 = 1$. Thus,

$$f_1 + f_2 + f_3 + \cdots + f_n = f_{n+2} - 1.$$

Now students can tackle the terms with odd indices. (Explain that the index tells *which* term, not the value of the term. Thus, in $f_9$, the index is 9 and it indicates the ninth term of the series—the ninth term is 34.) The sum $f_1 + f_3 + f_5 + f_7 + f_9 = 55$, the tenth term of the sequence, and $f_1 + f_3 + f_5 + f_7 + f_9 + f_{11} = f_{12} = 144$. Again, this can be shown algebraically for the sum of the first $n$ numbers with odd indices:

$$f_1 = f_2$$
$$f_3 = f_4 - f_2 \quad \text{(since } f_4 = f_2 + f_3)$$
$$f_5 = f_6 - f_4$$
$$\vdots$$
$$f_{2n-3} = f_{2n-2} - f_{2n-4}$$
$$f_{2n-1} = f_{2n} - f_{2n-2}.$$

By adding terms, $f_1 + f_3 + f_5 + \cdots + f_{2n-1} = f_{2n}$. Note that $n$ refers to the number of terms and not the index. Thus, for $n = 5$, the terms are $f_1, f_3, f_5, f_7, f_9$ and the index for the fifth term is 9, which in general notation is $2n - 1$.

The results are similar for terms with even indices. The sum $f_2 + f_4 + f_6 + f_8 = 33 = f_9 - 1$, and the sum $f_2 + f_4 + f_6 + f_8 + f_{10} = f_{11} - 1 = 88$.

If the algebraic derivations are too difficult for students, have them write more terms of the Fibonacci sequence and inductively reach conclusions. Continued testing of sums should convince them the relationships are true.

### Extension

If the extension is given to the entire class, some students may need guidance in the beginning. Have them try the sum of the squares of the first four Fibonacci numbers:

$$1^2 + 1^2 + 2^2 + 3^2 = 1 + 1 + 4 + 9 = 15 = 3 \cdot 5.$$

Thus, $f_1^2 + f_2^2 + f_3^2 = f_4 \cdot f_5$. Then they should experiment with other sums to see if the pattern checks. The general equation is $f_1^2 + f_2^2 + \cdots + f_n^2 = f_n \cdot f_{n+1}$. The algebraic derivation follows:

First note that for $k > 1$,

$$f_k f_{k+1} - f_{k-1} f_k = f_k(f_{k+1} - f_{k-1}) = f_k f_k = f_k^2.$$

Thus,

$$f_1^2 = f_1 f_2 - f_0 f_1 \quad \text{(where } f_0 = 0)$$
$$f_2^2 = f_2 f_3 - f_1 f_2$$
$$f_3^2 = f_3 f_4 - f_2 f_3$$
$$\vdots$$
$$f_{(n-1)}^2 = f_{n-1} f_n - f_{n-2} f_{n-1}.$$
$$f_n^2 = f_n f_{n+1} - f_{n-1} f_n.$$

By adding terms,

$$f_1^2 + f_2^2 + f_3^2 + \cdots + f_n^2 = f_n f_{n+1}.$$

18

# Pascal's Triangle

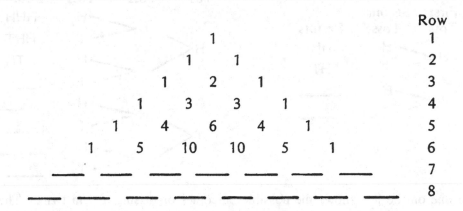

| | Row |
|---|---|
| 1 | 1 |
| 1   1 | 2 |
| 1   2   1 | 3 |
| 1   3   3   1 | 4 |
| 1   4   6   4   1 | 5 |
| 1   5   10   10   5   1 | 6 |
| __   __   __   __   __   __ | 7 |
| __   __   __   __   __   __   __   __ | 8 |

Find the numbers in the preceding diagram to form rows 7 and 8.

This arrangement of numbers is called *Pascal's triangle*. Various patterns of numbers from Pascal's triangle form other number sequences. Find the sum of each row of Pascal's triangle: 1, 2, 4, ____, ____, ____, ____, ____, ....

Now look at the diagonal row of numbers beginning 1, 3, 6, 10, .... What sequence of numbers do you get if you add each consecutive pair of these numbers? (That is, $1 + 3, 3 + 6, 6 + 10, ...$)?

---

Pascal's triangle can also be divided by diagonal lines as shown:

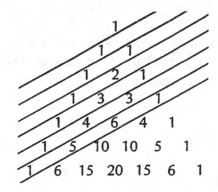

Find the sum of the numbers in each of these diagonal rows. What is the sequence of numbers called? _____

19

We can also use Pascal's triangle to find combinations of heads and tails when a coin is tossed several times. Look at the following tree diagrams:

The one on the left shows the possible outcomes of tossing a coin twice. The one on the right shows the outcomes of tossing it three times.

Fill in the blanks to show all the possible results. If a coin is tossed twice, how many ways can you get

two heads: _____ ; one head: _____ ; zero heads: _____ ?

If a coin is tossed three times, how many ways can you get

three heads: _____ ; two heads: _____ ;

one head: _____ ; zero heads: _____ ?

Suppose a coin were tossed four times. How many ways could you get two heads? _____

**EXTENSION!** Find these products: $(a + b)^0$, $(a + b)^1$, $(a + b)^2$, $(a + b)^3$, $(a + b)^4$. Write the products in descending powers of $a$. What do you notice about the coefficients? Find $(a + b)^6$ without multiplying.

# Teacher's Notes for Pascal's Triangle

*This activity provides an opportunity to explore many areas of mathematics—polygonal numbers, number theory in base 10 and other bases, sequences, probability, and binomial expansions. All these topics are related to Pascal's triangle.*

*Algebra skills are needed only in the Extension. However, students will find the activity most enjoyable if they have completed "The Fibonacci Sequence" and "Odds Are...."*

---

NCTM Standards

| 1 | 2 | 3 | 4 | 5 | 6 | 7 | 8 | 9 | 10 |
|---|---|---|---|---|---|---|---|---|----|
| ● | ● | ● |   |   |   |   |   | ● | ● |

---

### Presenting the Activity

Students should be able to find rows 7 and 8 in Pascal's triangle with no difficulty. If necessary, give them a hint by telling them to look at the two numbers above and slightly to the left and right of each number. Rows 7 and 8 are

```
  1   6  15  20  15   6   1
1   7  21  35  35  21   7   1
```

The sum of the numbers in each row gives the powers of 2: 1, 2, 4, 8, 16, 32, 64, 128, .... The numbers in the diagonal row 1, 3, 6, 10, ... are triangular numbers. Each number can be represented by a triangular array:

If students have completed "Triangular Numbers" in the *Making Pre-Algebra Come Alive* they will easily recognize this series of numbers.

When consecutive pairs of triangular numbers are added, perfect squares result: 1, 4, 9, 16, ....

The next pattern of sums produces the Fibonacci sequence: 1, 1, 2, 3, 5, 8, 13, ....

Students can find other patterns in Pascal's triangle. For example, the row number tells how many numbers there are in the row. Thus, row 5 has five numbers, row 6 has a six numbers, and so on. Also, the sum of the numbers in a diagonal is found in the next row:

Students may also recognize the powers of 11 in the first five rows of the triangle, and if they have studied the Extension to "Getting on Base," they will recognize 121 and 1331 as perfect squares and cubes in any base.

Next students consider coin tossing. The completed tree diagrams are

$$
\begin{array}{ll}
\textbf{Results} \\
H\!\!<\!\!\begin{array}{l}H\\T\end{array} & \begin{array}{l}HH\\HT\end{array}\\
T\!\!<\!\!\begin{array}{l}H\\T\end{array} & \begin{array}{l}TH\\TT\end{array}
\end{array}
\qquad
\begin{array}{ll}
& \textbf{Results}\\
H\!\!<\!\!\begin{array}{l}H\!\!<\!\!\begin{array}{l}H\\T\end{array}\\T\!\!<\!\!\begin{array}{l}H\\T\end{array}\end{array} & \begin{array}{l}HHH\\HHT\\HTH\\HTT\end{array}\\
T\!\!<\!\!\begin{array}{l}H\!\!<\!\!\begin{array}{l}H\\T\end{array}\\T\!\!<\!\!\begin{array}{l}H\\T\end{array}\end{array} & \begin{array}{l}THH\\THT\\TTH\\TTT\end{array}
\end{array}
$$

The results are filled in by simply following the lines in the tree diagram. Explain that the tree diagram gives all the possible results when a coin is tossed.

Now students count to find how many ways they can get different numbers of heads. For two tosses, the ways are

two heads: 1;     one head: 2;     zero heads: 1.

For three tosses,

three heads: 1;     two heads: 3;     one head: 3;     zero heads: 1.

Aha! These are rows in Pascal's triangle.

Thus, for four tosses, use the row 1   4   6   4   1. The "1" is for four heads, the "4" for three heads, the "6" for two heads, and so on. Therefore, if a coin is tossed four times, you could get two heads in six different ways.

If they have studied "Odds Are...," students will realize the application of Pascal's triangle to probability: The sum of the numbers in the row 1   4   6   4   1 is 16. This is the total number of outcomes when a coin is tossed four times. Each number in the row gives the number of various favorable outcomes. There is one way to get four heads, four ways to get three heads, six ways to get two heads, and so on. Thus, the probability of getting four heads when a coin is tossed four times is $\frac{1}{16}$, the probability of getting three heads is $\frac{4}{16}$, the probability of getting two heads is $\frac{6}{16}$, and so forth. Pascal's triangle applies for any *binomial* probability, a probability with only two possible outcomes. Thus, it is also applicable to finding the probabilities for families having a certain number of boys and girls.

## Extension

The Extension gives the algebraic application of Pascal's triangle. The binomial expansions are (the coefficient "1" is included to emphasize the pattern)

$$(a+b)^0 = 1,$$
$$(a+b)^1 = 1a + 1b,$$
$$(a+b)^2 = 1a^2 + 2ab + 1b^2,$$
$$(a+b)^3 = 1a^3 + 3a^2b + 3ab^2 + 1b^3,$$
$$(a+b)^4 = 1a^4 + 4a^3b + 6a^2b^2 + 4ab^3 + 1b^4.$$

The coefficients are rows of Pascal's triangle. The powers of $a$ descend in the same way that the number of heads descended in the coin tossing. To find $(a+b)^6$, students use row 7 of the triangle and write

$$a^6 + 6a^5b + 15a^4b^2 + 20a^3b^3 + 15a^2b^4 + 6ab^5 + b^6.$$

# Equations

- Algebraic Identities
- Patterns in Mathematics I
- Patterns in Mathematics II
- It's How You Use It... ★
- ...And How Much You Practice ★

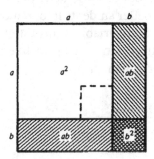

A major part of algebra courses is manipulating and solving equations. However, if equations are to be a truly useful tool in problem solving, students must have some idea of what the equations represent graphically and physically. All too often equation handling becomes a quasireligious incantation: Insert value of variable $A$ into slot $B$, push button $C$, turn crack $D$, and receive solution $E$. This leaves students uncomfortably vulnerable. They can be off by 3 orders of magnitude and suspect nothing is amiss. Students must develop some kind of gut feeling for what happens to one variable as the other is changed.

"Algebraic Identities" takes a step toward providing that intuition by considering such elementary examples as $(a + b)^2 = a^2 + 2ab + b^2$ and $(a - b)^2 = a^2 - 2ab + b^2$. Rather than ask students to memorize (and perhaps check by multiplication), students discover *visually* why these equations are valid. The technique described in this activity can be extended to other equations and operations throughout the normal study of high school algebra.

"Patterns in Mathematics I" can (and should) be used as soon as students have become acquainted with the simplest linear equations. It accomplishes two important facets of problem solving: First, it graphically illustrates the effects of the constant and coefficient in $y = ax + b$. Second, and more importantly, it shows students how to generalize from a table of observed data. As mentioned in the specific Teacher's Notes, you may find it desirable to first introduce the three "formulas" activities given in *Making Pre-Algebra Come Alive*.

"Patterns in Mathematics II" uses the same approach as "Patterns in Mathematics I," the difference being that quadratics are illustrated. Thus, the effects of altering the exponents are also shown. Both of these activities can be used with appropriate equations throughout both years of high school algebra.

At first glance, the last two activities in this section may seem to contradict what we've said in the preceding paragraphs, but only at first glance. These activities do present a step-by-step, somewhat mechanical method to solve certain equations. However, there are several good reasons for including them.

First, these activities treat very useful, rather special kinds of linear equations—Diophantine equations. Most standard texts ignore them, even though they can solve many realistic problems. Second, they provide a good challenge to your students' ingenuity. Finally, although the steps required to solve Diophantine equations are not individually difficult, students will have to read carefully and keep close tabs on their arithmetic.

Both of these activities give problems that are much more entertaining than your better students find in most standard texts. "It's How You Use It..." requires only one variable substitution to eliminate fractions. "...And How Much You Practice" requires up to five substitutions. This one is definitely only for your better students.

# Algebraic Identities

By now you can easily write expansions. Complete the following expansions:

$$(a + b)^2 = \underline{\hspace{8cm}},$$
$$a(b + c) = \underline{\hspace{8cm}},$$
$$(a - b)^2 = \underline{\hspace{8cm}}.$$

The expansions you have completed are examples of algebraic identities. You can show geometrically why they are true. Use the lengths given for $a$, $b$, and $c$ to form squares and rectangles that illustrate $a(b+c)$. Label the individual areas in your figure. Then use the given lengths to show geometrically the value of $(a + b)^2$. Again, label the individual areas of your figure.

$$\underline{\hspace{4cm}} \quad a$$
$$\underline{\hspace{2cm}} \quad b$$
$$\underline{\hspace{2.5cm}} \quad c$$

Finally, use the same method to represent $(a - b)^2$. Study your completed figure and tell why the expansion ends with the term $+ b^2$.

_____

_____

_____

_____

***EXTENSION!*** $a^2 + b^2 = c^2$ is not an algebraic identity, but it is one of the most basic and well-known theorems of geometry, the *Pythagorean theorem*. It states that the sum of the squares of the two shorter sides of a right triangle equals the square of the hypotenuse. This is illustrated by the diagram.

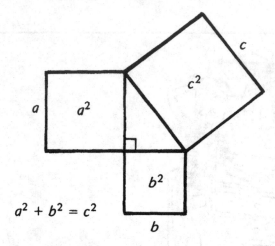

Use a separate sheet of paper to show by comparing areas why $a^2 + b^2 = c^2$.

# Teacher's Notes for Algebraic Identities

*Algebra and geometry are too often put into separate categories and treated as though there were little or no connection between them. This is unfortunate; not only are the relationships between algebra and geometry intrinsically interesting, they are mutually supportive. That is, they can provide concrete, graphic examples that make abstractions much easier to grasp. This activity provides one such example.*

*Students who are having some rough sledding with algebra should feel comfortable with this activity and may profit even further by using "Geometric Dissections" in **Making Pre-Algebra Come Alive** as a warmup. Students who are especially pleased by the insights of this activity will enjoy exploring, "Constructing Segments" in the **Making Geometry Come Alive** which extends the concepts presented here and lays the groundwork for a consideration of vectors.*

### NCTM Standards

| 1 | 2 | 3 | 4 | 5 | 6 | 7 | 8 | 9 | 10 |
|---|---|---|---|---|---|---|---|---|----|
|   | • | • |   |   |   |   |   | • | •  |

### Presenting the Activity

Your students will have no difficulty completing the expansions.

$$(a + b)^2 = a^2 + 2ab + b^2,$$
$$a(b + c) = ab + ac,$$
$$(a - b)^2 = a^2 - 2ab + b^2.$$

(If they do have difficulty, this activity should be deferred.)

The first exercise will be obvious to students once they realize what is being asked for. We've included this very easy example here simply to make this point. Work through this question on the chalkboard. Students should realize that the line separating areas $ab$ and $ac$ may be either horizontal or vertical.

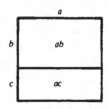

The next figure should be as shown by

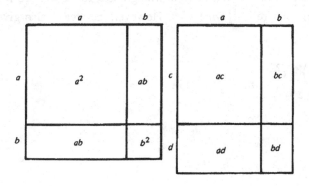

27

The square is composed of four quadrilaterals (two squares and two rectangles) whose areas correspond to the terms in the expansion. If this idea needs to be reinforced, you may want to illustrate $(a+b)(c+d)$, where $a$, $b$, $c$, and $d$ all have different magnitudes. Four rectangles are formed, but the pattern is similar, and students can easily see that $(a+b)^2$ is a special case of $(a+b)(c+d)$. (A rigorous proof is given in Euclid's *Elements*, Book II, Proposition 4.)

The last question on the student page (prior to the Extension) should provoke the most student interest. The two areas $ab$ to be subtracted are shaded in the figure

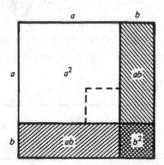

In subtracting $ab$ twice, however, area $b^2$ is removed twice. Thus, if the term $b^2$ were not added in the expansion, the area represented by $a^2 - 2ab$ gives only the area that omits what is bounded by the dotted line as well as the shaded area.

### Extension

The Extension gives students an opportunity to devise one of the hundreds of proofs of the Pythagorean theorem. In most cases, it's wise to give first-year algebra students a hint. To get them started, draw the following figure on the chalkboard:

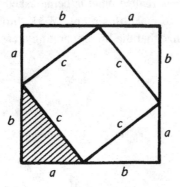

Students already know that the area of the shaded right triangle is $\frac{1}{2}ab$. There are four of these, so the area of the large square is $4(\frac{1}{2}ab) + c^2$. However, the area is also given by $(a+b)^2$,

$$(a+b)^2 = 4\left(\frac{1}{2}ab\right) + c^2,$$
$$a^2 + 2ab + b^2 = 2ab + c^2,$$
$$a^2 + b^2 = c^2.$$

# Patterns in Mathematics I

| | (a) | | | (b) | | | (c) | | | (d) | |
|---|---|---|---|---|---|---|---|---|---|---|---|
| $x$ | $y$ | $D$ | $x$ | $y$ | $D$ | $x$ | $y$ | $D$ | $x$ | $y$ | $D$ |
| 0 | 0 | | 0 | 0 | | 0 | 1 | | 0 | 3 | |
| 1 | 2 | __ | 1 | 3 | __ | 1 | 1.5 | __ | 1 | 5 | __ |
| 2 | 4 | __ | 2 | 6 | __ | 2 | 2.0 | __ | 2 | 7 | __ |
| 3 | 6 | __ | 3 | 9 | __ | 3 | 2.5 | __ | 3 | __ | __ |
| 4 | __ | __ | 4 | __ | __ | 4 | __ | __ | 4 | __ | __ |
| 5 | __ | __ | 5 | __ | __ | 5 | __ | __ | 5 | __ | __ |

Look at Tables (a), (b), (c), and (d). You should see patterns developing in the ways the values of $y$ grow. Fill in the missing values of $y$.

Now look at the columns headed by $D$. $D$ stands for the difference between a given value of $y$ and the preceding value of $y$. Thus, in Table (a), when $x$ increases from 1 to 2, $y$ goes from _____ to _____ .

The difference $D =$ _____ . In Table (b), if we add 1 to $x$ , $y$ goes up by _____ . In Table (c), $y$ goes up by _____ when we add 1 to $x$, and in Table (d), it goes up by _____ .

In Tables (a) and (b), when $x = 0$, $y =$ _____ . From then on, whatever $x$ equals, $y$ equals _____ in Table (a) and _____ in Table (b).

Now you're in a position to make general statements of $y$ in terms of $x$; in other words, equations.

     For Table (a), $y =$ _____ $x$.
     For Table (b), $y =$ _____ .

In Table (c), when $x = 0$, $y =$ _____ , and in Table (d), $y =$ _____ when $x = 0$. So in Tables (c) and (d), $y$ has a "head start" on $x$. This head start is shown by a *constant* in their equations.

     For Table (c), $y =$ _____ $x +$ _____ .
     For Table (d), $y =$ _____ $x +$ _____ .

Now the general pattern becomes clear. In the equations that describe all four of the tables, the coefficient of $x$ is given by _____ .

The constant is shown by the value of _____ when _____ $= 0$.

Now show the points for the different values of $x$ and $y$ from Table (a). Then join them together to form the graph of the equation $y = 2x$.

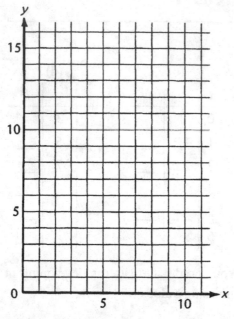

Do this for Tables (b), (c), and (d) as well. What do these graphs tell you about the effect of the coefficient on the graph? _____

_____

What do they tell you about the effect of the constant? _____

_____

_____

***EXTENSION!*** Use the graphs on the attached sheet to find (a) the equations for the graphs and (b) tables that show the same information as the graphs.

# Teacher's Notes for Patterns in Mathematics I

*This activity gets to the heart of scientific and economic problem solving: tabulating empirical data and extracting from that data a general statement of how the variables are related. It's one thing to be able to solve an equation that is given to you, but a researcher must often start with raw data and then state an appropriate equation. This is a much more sophisticated skill and accounts, in large part, for the difficulties many students have with word problems.*

*This is very much in accord with what is said in the three "Writing Formulas" activities in **Making Pre-Algebra Come Alive**: Mathematics must be treated as a language when it is used for empirical problem solving. You may wish to present one or more of the activities from **Making Pre-Algebra Come Alive** either as preparation or as reinforcement for this one. Before attempting this activity, students should be familiar with simple linear equations of the form $y = ax + b$.*

---------------- NCTM Standards ----------------

| 1 | 2 | 3 | 4 | 5 | 6 | 7 | 8 | 9 | 10 |
|---|---|---|---|---|---|---|---|---|----|
| • | • | • | | | • | | | • | • |

## Presenting the Activity

Most students will have no difficulty completing Tables (a), (b), (c), and (d) as shown

| (a) | | | (b) | | | (c) | | | (d) | | |
|---|---|---|---|---|---|---|---|---|---|---|---|
| $x$ | $y$ | $D$ | $x$ | $y$ | $D$ | $x$ | $y$ | $D$ | $x$ | $y$ | $D$ |
| 0 | 0 | | 0 | 0 | | 0 | 1 | | 0 | 3 | |
| 1 | 2 | 2 | 1 | 3 | 3 | 1 | 1.5 | 0.5 | 1 | 5 | 2 |
| 2 | 4 | 2 | 2 | 6 | 3 | 2 | 2.0 | 0.5 | 2 | 7 | 2 |
| 3 | 6 | 2 | 3 | 9 | 3 | 3 | 2.5 | 0.5 | 3 | 9 | 2 |
| 4 | 8 | 2 | 4 | 12 | 3 | 4 | 3.0 | 0.5 | 4 | 11 | 2 |
| 5 | 10 | 2 | 5 | 15 | 3 | 5 | 3.5 | 0.5 | 5 | 13 | 2 |
| $y = 2x$ | | | $y = 3x$ | | | $y = \frac{1}{2}x + 1$ | | | $y = 2x + 3$ | | |

However, to assure that all students are on track, it is best to walk them through the questions in paragraph 2. It is from there clear that $D$ defines the coefficient of $x$ and that the constant is the value of $y$ when $x$ equals zero. Have the students substitute different values of $x$ in their equations to make sure that their derived values of $y$ agree with those in the tables.

Whereas Tables (a) and (b) focus only on the derivation of the coefficient of $x$ or slope of the equation, Tables (c) and (d) introduce the effect and significance of the constant in $y = ax + b$. After completing the graphs as shown in the accompanying diagram students can see the effect of $b$, the point at which the graph crosses the $y$ axis and that we have called a "head start."

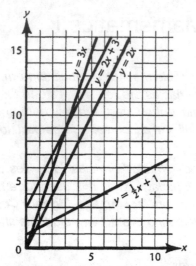

It's always a good practice to illustrate ideas such as this with examples taken from situations with which the students are already familiar. Such an example can be provided by comparing the balances of two savings accounts. Assume that on January 1 Amy has $70 in her account and Bill has $20 in his. If, over the next four months they deposit $5 and $10 per week, respectively, the accounts (ignoring interest) can be represented as

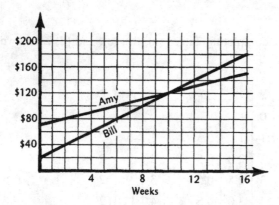

Although Amy's graph has a $50 head start on Bill's graph, Bill's graph overtakes Amy's after 10 weeks because the slope of Bill's graph is double that of Amy's.

### Extension

The tables and equations for the graphs are

| | (1) | | | | (2) | | | | (3) | | | | (4) | | |
|---|---|---|---|---|---|---|---|---|---|---|---|---|---|---|---|
| $x$ | $y$ | $D$ | | $x$ | $y$ | $D$ | | $x$ | $y$ | $D$ | | $x$ | $y$ | $D$ |
| $-5$ | 0 | | | $-5$ | 0 | | | 0 | $-3$ | | | $-2$ | 0 | |
| 0 | 5 | | | 0 | 2 | | | 1 | $-1$ | 2 | | 0 | $-1$ | |
| 1 | 6 | 1 | | 1 | $2\frac{1}{2}$ | $\frac{1}{3}$ | | 2 | 1 | 2 | | 1 | $-1.5$ | $-0.5$ |
| 2 | 7 | 1 | | 2 | $2\frac{2}{3}$ | $\frac{1}{3}$ | | 3 | 3 | 2 | | 2 | $-2$ | $-0.5$ |
| 3 | 8 | 1 | | 3 | 3 | $\frac{1}{3}$ | | 4 | 5 | 2 | | 3 | $-2.5$ | $-0.5$ |
| $y = x + 5$ | | | | $y = \frac{1}{3}x + 2$ | | | | $y = 2x - 3$ | | | | $y = \frac{1}{2}x - 1$ | | |

Note that negative coefficients and constants are introduced in Tables (3) and (4). Ask your students to think of practical situations that would reflect these negative elements.

32

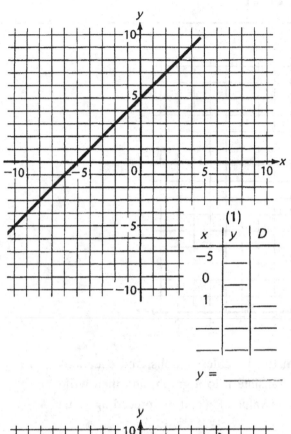

**(1)**

| x | y | D |
|---|---|---|
| −5 | | |
| 0 | | |
| 1 | | |
| | | |
| | | |

y = _____

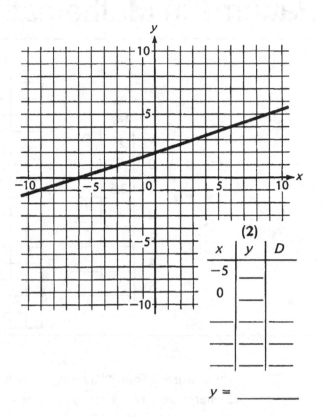

**(2)**

| x | y | D |
|---|---|---|
| −5 | | |
| 0 | | |
| | | |
| | | |

y = _____

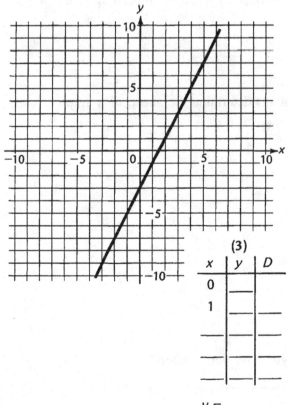

**(3)**

| x | y | D |
|---|---|---|
| 0 | | |
| 1 | | |
| | | |
| | | |
| | | |

y = _____

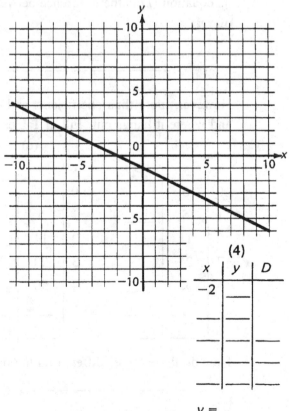

**(4)**

| x | y | D |
|---|---|---|
| −2 | | |
| | | |
| | | |
| | | |
| | | |

y = _____

# Patterns in Mathematics II

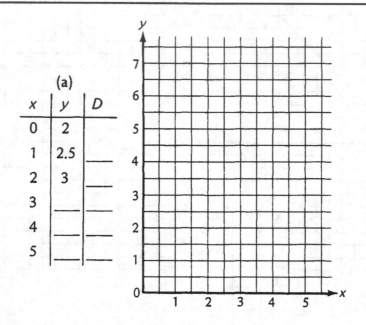

**(a)**

| x | y | D |
|---|-----|---|
| 0 | 2 | |
| 1 | 2.5 | ___ |
| 2 | 3 | ___ |
| 3 | ___ | ___ |
| 4 | ___ | ___ |
| 5 | ___ | ___ |

As you learned from "Patterns in Mathematics I" tables, graphs, and equations are interchangeable. Complete Table (a), translate it to a graph, and then write its equation (D is the difference between one value of y and the preceding value of y):

$$y = \underline{\hspace{5cm}}.$$

However, more than coefficients and constants determine the shape of a graph and where it hangs out. Fill in Tables (b) and (c):

**(b)**

| x | y | D |
|---|-----|---|
| 0 | 3 | |
| 1 | 5 | ___ |
| 2 | 11 | ___ |
| 3 | 21 | ___ |
| 4 | 35 | ___ |
| 5 | ___ | ___ |
| 6 | ___ | ___ |

**(c)**

| x | y | D |
|---|-----|---|
| 0 | 2 | |
| 1 | 5 | ___ |
| 2 | 14 | ___ |
| 3 | 29 | ___ |
| 4 | 50 | ___ |
| 5 | ___ | ___ |
| 6 | ___ | ___ |

How do these tables differ from the ones you have done before? _____

_____

_____

Try adding another column to Tables (b) and (c). Head this column with $D'$, where $D'$ equals the difference between successive $D$s (just as $D$ shows the difference between successive $y$s). Does $D'$ remain constant? _____

The fact that we have to go to a $D'$ term before reaching a constant tells us we will have an $x^2$ term in the equation. Now try to write equations for Tables (b) and (c).

For Table (b), $y =$ _____ .

For Table (c), $y =$ _____ .

How do the coefficients of $x^2$ compare with $D'$? _____

Substitute different values of $x$ into your equations. Do your values for $y$ agree with those in the tables? _____

Suppose you have a pattern of circles such as

Let $x$ equal the number of circles in a given row and let $y$ equal the total number of the circles in row $x$ plus the circles in the smaller rows. Fill in the following table.

| Row ($x$) | Total Circles ($y$) | $D$ | $D'$ |
|-----------|---------------------|-----|------|
| 1 | 1 | | |
| 2 | 3 | ___ | |
| 3 | 6 | ___ | ___ |
| 4 | ___ | ___ | ___ |
| 5 | ___ | ___ | ___ |
| 6 | ___ | ___ | ___ |

What equation describes the tables you have made? _____

**EXTENSION!** Use the graphs on the attached sheet to find (a) the equations for the graphs and (b) tables that show the same information as the graphs.

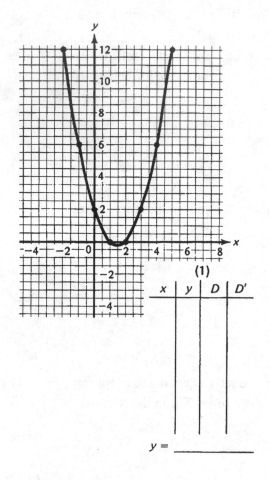

(1)

| x | y | D | D' |
|---|---|---|---|
|   |   |   |    |

y = _____

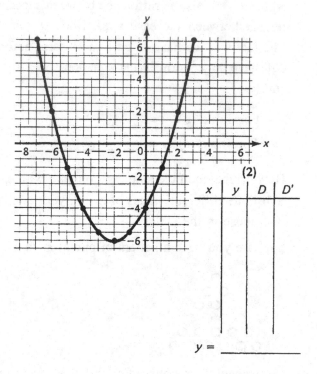

(2)

| x | y | D | D' |
|---|---|---|---|
|   |   |   |    |

y = _____

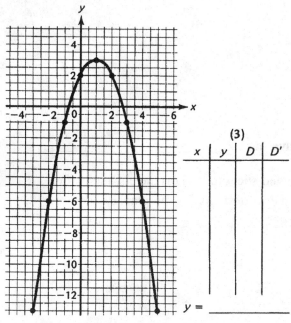

(3)

| x | y | D | D' |
|---|---|---|---|
|   |   |   |    |

y = _____

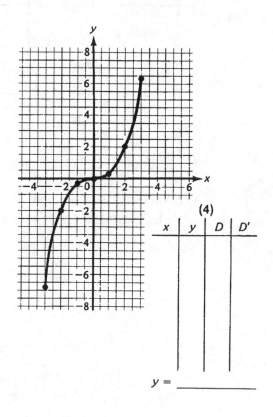

(4)

| x | y | D | D' |
|---|---|---|---|
|   |   |   |    |

y = _____

36

# Teacher's Notes for Patterns in Mathematics II

*This activity was initially combined with "Patterns in Mathematics I." During field testing, however, we quickly learned that students who profited from the examination of linear equations had excessive difficulties with the second-power equations. Accordingly, it is recommended that Part I be presented during the first semester, and Part II deferred until quadratics have been introduced. Alternatively, both are well suited for presentation near the beginning of second-year algebra courses.*

| | | | | NCTM Standards | | | | | |
|---|---|---|---|---|---|---|---|---|---|
| 1 | 2 | 3 | 4 | 5 | 6 | 7 | 8 | 9 | 10 |
| • | • | • | | | • | | | • | • |

### Presenting the Activity

Students will have no difficulty completing Table (a) and drawing a graph as shown:

(a)

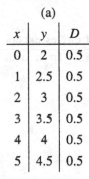

| x | y | D |
|---|---|---|
| 0 | 2 | 0.5 |
| 1 | 2.5 | 0.5 |
| 2 | 3 | 0.5 |
| 3 | 3.5 | 0.5 |
| 4 | 4 | 0.5 |
| 5 | 4.5 | 0.5 |

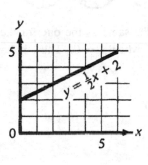

Review the effects of the constant and the coefficient on the shape and location of the graph. Also note that for $y = ax + b$, $D = a$.

When students fill in Tables (b) and (c), they'll notice that $D$ does not remain constant. If they add another column to the tables, however, they find this second difference, $D'$, does yield a constant.

| | (b) | | | | | (c) | | |
|---|---|---|---|---|---|---|---|---|
| x | y | D | D' | | x | y | D | D' |
| 0 | 3 | | | | 0 | 2 | | |
| 1 | 5 | 2 | | | 1 | 5 | 3 | |
| 2 | 11 | 6 | 4 | | 2 | 14 | 9 | 6 |
| 3 | 21 | 10 | 4 | | 3 | 29 | 15 | 6 |
| 4 | 35 | 14 | 4 | | 4 | 50 | 21 | 6 |
| 5 | 53 | 18 | 4 | | 5 | 77 | 27 | 6 |
| 6 | 75 | 22 | 4 | | 6 | 110 | 33 | 6 |

$$y = 2x^2 + 3 \qquad\qquad y = 3x^2 + 2$$

Some students will use the analogy from Table (a) (and from "Patterns in Mathematics I") and decide that Table (b) yields $y = 4x^2 + 3$ and Table (c) yields $y = 6x^2 + 2$. When they try substituting, however, these equations don't hold up. The value of $y$ at $x = 0$

37

still yields the constant, but $D'$ represents twice the coefficient of $x^2$. You can easily demonstrate this with the preceding table or, in general form, with the table

| $x$ | $y$ | $D$ | $D'$ |
|---|---|---|---|
| 0 | $c$ | | |
| 1 | $a+b+c$ | $a+b$ | |
| 2 | $4a+2b+c$ | $3a+b$ | $2a$ |
| 3 | $9a+3b+c$ | $5a+b$ | $2a$ |
| 4 | $16a+4b+c$ | $7a+b$ | $2a$ |

that yields $y = ax^2 + bx + c$.

Deriving the equation for the rows of circles can give beginning algebra students some trouble at first. For this reason, it is usually advisable to present another, similar example such as: What is the *largest* number of pieces you can make by cutting a circle with $x$ cuts?

The table for this problem is exactly the same as the one for the rows of circles except that there is a constant: $y = 1$, not 0, when $x = 0$. Thus, if you subtract 1 from each value of $y$ in the table

| $x$<br>No. of Cuts | $y$<br>No. of Pieces | $D$ | $D'$ |
|---|---|---|---|
| 0 | 1 | | |
| 1 | 2 | 1 | |
| 2 | 4 | 2 | 1 |
| 3 | 7 | 3 | 1 |
| 4 | 11 | 4 | 1 |
| 5 | 16 | 5 | 1 |

$$y = \tfrac{1}{2}x^2 + \tfrac{1}{2}x + 1$$

you have the table for the rows of circles.

Examine the pattern. The second difference is $2a$, twice the value of the coefficient of $x^2$. The value of the first difference when $x = 1$ is $a + b$. Because we know the value of $a$ (it is one-half of the second difference), we can find the value of $b$ by subtracting $a$ from the first difference $(a + b)$. The constant is the value of $y$ when $x$ is zero. Therefore, the constant is 1. $D'$ is $2a$. Because $D'$ is 1, the value of $a$ is $\tfrac{1}{2}$. $D$ is $a + b$. Because $D$ is 1 and $a$ is $\tfrac{1}{2}$, the value of $b$ is $\tfrac{1}{2}$. The equation therefore, is $y = \tfrac{1}{2}x^2 + \tfrac{1}{2}x + 1$.

## Extension

The tables and equations for the graphs are:

### (1)

| $x$ | $y$ | $D$ | $D'$ |
|---|---|---|---|
| −2 | 12 | | |
| −1 | 6 | −6 | |
| 0 | 2 | −4 | 2 |
| 1 | 0 | −2 | 2 |
| 2 | 0 | 0 | 2 |
| 3 | 2 | 2 | 2 |
| 4 | 6 | 4 | 2 |
| 5 | 12 | 6 | 2 |

$$y = x^2 - 3x + 2$$

### (2)

| $x$ | $y$ | $D$ | $D'$ |
|---|---|---|---|
| −7 | 6.5 | | |
| −6 | 2 | −4.5 | |
| −5 | −1.5 | −3.5 | 1 |
| −4 | −4 | −2.5 | 1 |
| −3 | −5.5 | −1.5 | 1 |
| −2 | −6 | −0.5 | 1 |
| −1 | −5.5 | 0.5 | 1 |
| 0 | −4 | 1.5 | 1 |
| 1 | −1.5 | 2.5 | 1 |
| 2 | 2 | 3.5 | 1 |
| 3 | 6.5 | 4.5 | 1 |

$$y = \tfrac{1}{2}x^2 + 2x - 4$$

### (3)

| $x$ | $y$ | $D$ | $D'$ |
|---|---|---|---|
| −3 | −13 | | |
| −2 | −6 | 7 | |
| −1 | −1 | 5 | −2 |
| 0 | 2 | 3 | −2 |
| 1 | 3 | 1 | −2 |
| 2 | 2 | −1 | −2 |
| 3 | −1 | −3 | −2 |
| 4 | −6 | −5 | −2 |
| 5 | −13 | −7 | −2 |

$$y = -x^2 + 2x + 2$$

### (4)

| $x$ | $y$ | $D$ | $D'$ | $D''$ |
|---|---|---|---|---|
| −3 | −6.75 | | | |
| −2 | −2 | 4.75 | | |
| −1 | −0.25 | 1.75 | −3 | |
| 0 | 0 | 0.25 | −1.5 | 1.5 |
| 1 | 0.25 | 0.25 | 0 | 1.5 |
| 2 | 2 | 1.75 | 1.5 | 1.5 |
| 3 | 6.75 | 4.75 | 3 | 1.5 |

$$y = \tfrac{1}{4}x^3$$

Note that for Table (4), $D'$ does not yield a constant. $D''$ does, and so the equation is cubic.

# It's How You Use It...

An office clerk spent $12.50 on 20¢ stamps and 15¢ postcards. How many stamps and how many postcards were porchased?

Let $x$ equal the number of stamps and let $y$ equal the number of postcards.

Then $20x + 15y = 1250$ or $4x +$ _____ = _____ .

What kind of numbers must $x$ and $y$ be? _____

You could probably find a solution for this equation just by trying different values for $x$ and $y$, but there's a way to find *all* the whole number solutions. First, solve for the variable whose coefficient has the smaller absolute value. In this problem, solve for $y$:

$$4x + 3y = 250$$
$$3y = 250 - 4x$$
$$y = \frac{250}{3} - \frac{4}{3}x$$
$$y = 83\frac{1}{3} - 1\frac{1}{3}x.$$

Separate the whole numbers and the fractions:

$$y = 83 + \frac{1}{3} - x - \frac{x}{3}$$
$$= 83 - x + \frac{1-x}{3}.$$

Then use a new variable to represent the fraction. That is, let $t = \frac{1-x}{3}$. Solve this equation for $x$:

$$x = \underline{\hspace{4cm}}.$$

40

Now substitute this value of $x$ in the original equation and solve for $y$: $y =$
_____ .

For various values of $t$, you can find values of $x$ and $y$. To find all the whole number solutions, remember $x > 0$ and $y > 0$. This means

$$1 - 3t > 0 \quad \text{and} \quad 82 + 4t > 0,$$
$$3t > 1 \quad \text{and} \quad 4t > -82,$$
$$t > \frac{1}{3} \quad \text{and} \quad t > -20\frac{1}{2}.$$

What possible values can $t$ have? _____

How many whole number solutions are there for $4x + 3y = 250$? _____

Try this problem: Two boxes and 11 bags weigh 35 lb. If no fractions or decimals are allowed, how much must each box and bag weigh? (Find all solutions.)

Let $x$ equal the number of boxes and let $y$ equal the number of bags. What is the equation? _____

Solve for $x$:

$\qquad x =$ _____ .

Let $t$ equal the fraction and solve for $y$:

$\qquad y =$ _____ .

Now substitute this value of $y$ in the original equation and solve for $x$:

$\qquad x =$ _____ .

What possible values can $t$ have? _____

What are the whole number solutions? _____

***EXTENSION!*** The villagers of the romantic Aspirin Islands play a game called *forgle*. Points are scored by getting flamostats (passing the uhlich over the rebaz). Points are lost when hiccups are made (making the gingling come to rest inside the inkwell). In a recent match, the Obese Parakeets made 7 flamostats and 3 hiccups for a total of 23 points. How many points could flamostats and hiccups be worth?

# Teacher's Notes for It's How You Use It...

When confronted by a single linear equation containing two variables, students who have been exposed to solving systems of simultaneous equations quickly conclude: "We don't have enough information." However, as Leonard Euler found out, it ain't just what you got; it's how you use it. This activity shows how to apply Euler's method to realistic Diophantine equations. "... And How Much You Practice" extends the techniques presented here.

――――――――――――――――― NCTM Standards ―――――――――――――――

| 1 | 2 | 3 | 4 | 5 | 6 | 7 | 8 | 9 | 10 |
|---|---|---|---|---|---|---|---|---|----|
| • | • |   |   |   | • |   |   |   |    |

### Presenting the Activity

Discuss the first problem to be sure students understand how the problem is translated to an equation. Because the problem deals with stamps and postcards, $x$ and $y$ must be whole numbers.

Equations of the form $ax + by = k$ are called *Diophantine equations* if their solutions must be integers. Diophantine equations will have an infinite number of integral solutions if the greatest common factor of $a$ and $b$ is also a factor of $k$. For example, $3x - 18y = 40$ will not have integral solutions because 3 is not a factor of 40. For $4x + 3y = 250$, the greatest common factor of 4 and 3 is 1, which is also a factor of 250. Thus, there are an infinite number of integral solutions.

The method shown on the student page for finding integral solutions was developed by Euler. Although the method isn't difficult, it will be new to students, so it's a good idea to work through the first problem carefully. Solving the equation in $t$ for $x$,

$$t = \frac{1 - x}{3},$$
$$3t = 1 - x,$$
$$x = 1 - 3t.$$

Now this value of $x$ is substituted in the original equation and the students solve for $y$:

$$4x + 3y = 250,$$
$$4(1 - 3t) + 3y = 250,$$
$$4 - 12t + 3y = 250,$$
$$-12t + 3y = 246,$$
$$3y = 246 + 12t,$$
$$y = 82 + 4t.$$

Because we are only interested in positive integral values for $x$ and $y$, we limit the values of $t$ as shown on the student page. Thus, $t$ can equal $0, -1, -2, \ldots, -20$ and there are 21 whole number solutions for $4x + 3y = 250$.

In the next problem, students translate to the equation $2x + 11y = 35$ and then solve for $x$:

$$2x + 11y = 35,$$
$$2x = 35 - 11y,$$
$$x = 17\frac{1}{2} - 5\frac{1}{2}y,$$
$$x = 17 + \frac{1}{2} - 5y - \frac{y}{2},$$
$$x = 17 - 5y + \frac{1 - y}{2}.$$

Now let $t = \frac{1-y}{2}$ and solve for $y$:

$$t = \frac{1 - y}{2},$$
$$2t = 1 - y,$$
$$y = 1 - 2t.$$

Substitute in the original equation and solve for $x$:

$$2x + 11y = 35,$$
$$2x + 11(1 - 2t) = 35,$$
$$2x + 11 - 22t = 35,$$
$$2x - 22t = 24,$$
$$2x = 24 + 22t,$$
$$x = 12 + 11t.$$

Because we are only interested in positive values of $x$ and $y$, we limit $t$ as in the first problem. We find $t < \frac{1}{2}$ and $t > -1\frac{1}{11}$. Thus, there are two possible values of $t$ that will give positive integers for $x$ and $y$: $t = 0, -1$. For $t = 0$, $x = 12$ and $y = 1$. For $t = -1$, $x = 1$ and $y = 3$.

Note that all the problems in this activity require the introduction of only one new variable to eliminate the fractions.

### Extension

This problem translates to $7x - 3y = 23$. This equation is then solved for $y$:

$$7x - 3y = 23,$$
$$-3y = 23 - 7x,$$
$$y = -7\frac{2}{3} + 2\frac{1}{3}x,$$
$$y = -7 - \frac{2}{3} + 2x + \frac{x}{3},$$
$$y = 2x - 7 + \frac{x - 2}{3}.$$

(Note that the absolute values of the coefficients are used to determine which variable to solve for. Thus, if the equation were $3x - 7y = 23$, we would solve for $x$).

43

Let $t = \frac{x-2}{3}$ and solve for $x$:

$$t = \frac{x-2}{3},$$
$$3t = x - 2,$$
$$x = 3t + 2.$$

Substitute in the original equation and solve for $y$:

$$7x - 3y = 23,$$
$$7(3t + 2) - 3y = 23,$$
$$21t + 14 - 3y = 23$$
$$21t - 3y = 9,$$
$$-3y = 9 - 21t,$$
$$y = -3 + 7t.$$

Then $t$ is limited, so only positive values of $x$ and $y$ are found. We find $t > -\frac{2}{3}$ and $t > \frac{3}{7}$. Thus, any integral value of $t > 0$ will produce positive values of $x$ and $y$ and there are an infinite number of solutions.

# ...And How Much You Practice

"EIGHT FOR DINNER, PLEASE!"

In "It's How You Use It...," you found the solutions to an equation in $x$ and $y$ by introducing a new variable $t$. By introducing $t$, you created an equation with no fractions in it. Sometimes it's necessary to introduce more variables before you get an equation without any fractions.

Consider this problem: The German Club bought five boxes of lemon cupcakes and nine boxes of chocolate cupcakes for a party. At the party, all 187 cupcakes were eaten. How many of each kind of cupcake were there per box?

The equation is $5x + 9y = 187$. First solve for $x$, separating the whole numbers and the fractions:

$$5x + 9y = 187,$$
$$5x = 187 - 9y,$$
$$x = 37\frac{2}{5} - 1\frac{4}{5}y,$$
$$x = 37 + \frac{2}{5} - y - \frac{4}{5}y,$$
$$x = 37 - y - \frac{4y - 2}{5}.$$

Let $t = \frac{4y-2}{5}$ and solve for $y$:

$$t = \frac{4y - 2}{5},$$
$$5t = 4y - 2,$$
$$4y = 5t + 2,$$
$$y = 1\frac{1}{4}t + \frac{2}{4},$$
$$y = t + \frac{t + 2}{4}.$$

Then use a new variable to represent the fraction in this equation. That is, let $u = \frac{t+2}{4}$. Solve this equation for $t$:

$t =$ _____ .

Substitute this value of $t$ in the equation for $t$ and solve for $y$:

$y =$ _____ .

Now substitute this value of $y$ in the original equation and solve for $x$:

$x =$ _____ .

For various values of $u$, you can find values of $x$ and $y$. Because $x$ and $y$ must be positive, what possible values can $u$ have? _____
What are the whole number solutions? _____

Try this problem: Five 4-H Club members spent an afternoon planting tomato plants. Each of the five set the same number of plants. During the night, eight rabbits dropped by the 4-H garden for dinner. Each rabbit ate the same number of plants. In the morning 39 uneaten plants remained. How many plants did each 4-H Club member plant and how many did each rabbit eat? _____

***EXTENSION!*** A government official bought some bags of jelly beans to take along on an airplane ride to Washington. The red jelly beans cost 13¢ per bag and the black ones cost 21¢. If the official spent $2.61, how many bags of each kind were purchased?

# Teacher's Notes for ...And How Much You Practice

*In "It's How You Use It...," students had to introduce only one new variable to solve the original equation. Life is generally not so kind; two or more steps are usually required to eliminate all the fractions so that integral solutions are easily obtained. This activity extends the techniques of Euler to include two, three, and five substitution steps.*

*Students should successfully complete "It's How You Use It..." before working on this activity.*

---

NCTM Standards

| 1 | 2 | 3 | 4 | 5 | 6 | 7 | 8 | 9 | 10 |
|---|---|---|---|---|---|---|---|---|----|
| • | • |   |   |   | • |   |   |   |    |

---

### Presenting the Activity

The first problem requires the introduction of two new variables. Although the process is the same, it's a good idea to work through the problem carefully. Solving the equation in $u$ for $t$ gives

$$u = \frac{t+2}{4},$$
$$4u = t + 2,$$
$$t = 4u - 2.$$

Now substitutions are made in reverse order. First the value of $t$ is substituted in the equation for $t$ and students solve for $y$:

$$4u - 2 = \frac{4y-2}{5},$$
$$20u - 10 = 4y - 2,$$
$$20u - 8 = 4y,$$
$$y = 5u - 2.$$

Then this value of $y$ is substituted in the original equation and students solve for $x$:

$$5x + 9y = 187,$$
$$5x + 9(5u - 2) = 187,$$
$$5x + 45u - 18 = 187,$$
$$5x + 45u = 205,$$
$$5x = 205 - 45u,$$
$$x = 41 - 9u.$$

Thus, for various integral values of $u$, integral values of $x$ and $y$ can be found. Because $x$ and $y$ must be positive, $5u - 2 > 0$ and $41 - 9u > 0$, or $u > \frac{2}{5}$ and $u > 4\frac{5}{9}$. The only values of $u$ that will yield whole numbers for $x$ and $y$ are $u = 1, 2, 3, 4$, so the solutions are $x = 32$, $y = 3$; $x = 23$, $y = 8$; $x = 14$, $y = 13$; $x = 5$, $y = 18$.

To give students additional practice with equations that require the introduction of two variables, use $7x - 31y = 2$ and $18x - 53y = 3$. Both have an infinite number of positive integral solutions.

In the next problem, students must introduce three new variables. The equation is $5x - 8y = 39$. (Note that the equation is solved for the variable that has the coefficient with the smallest absolute value.) The steps are as follows

$$5x - 8y = 39,$$
$$5x = 39 + 8y,$$
$$x = 7\frac{4}{5} + 1\frac{3}{5}y,$$
$$x = 7 + y\frac{4 + 3y}{5}.$$

Let $t = \frac{4+3y}{5}$ and solve for $y$:

$$5t = 4 + 3y,$$
$$3y = 5t - 4,$$
$$y = 1\frac{2}{3}t - 1\frac{1}{3},$$
$$y = t - 1 + \frac{2t - 1}{3}.$$

Let $u = \frac{2t-1}{3}$ and solve for $t$:

$$3u = 2t - 1,$$
$$2t = 3u + 1,$$
$$t = 1\frac{1}{2}u + \frac{1}{2},$$
$$t = u + \frac{u + 1}{2}.$$

Let $v = \frac{u+1}{2}$ and solve for $u$:

$$2v = u + 1,$$
$$u = 2v - 1.$$

Substituting in reverse order gives $t = 3v - 1$, $y = 5v - 3$, and $x = 3 + 8v$.

To find the whole number solutions, $5v - 3 > 0$ and $3 + 8v > 0$ or $v > \frac{3}{5}$, and $v > -\frac{3}{8}$. Thus, $v = 1, 2, 3, \ldots$ will yield whole number values for $x$ and $y$ and there are an infinite number of solutions. Have students find a few solutions and decide which is the most reasonable for the problem.

### Extension

This problem requires introducing five new variables. The problem translates to $13x + 21y = 261$. Solving for $x$ gives

$$x = 20 - y - \frac{8y - 1}{13}.$$

Letting $t$ equal the fraction and solving for $y$ gives

$$y = t + \frac{5t + 1}{8}.$$

Letting $u$ equal the fraction and solving for $t$ gives

$$t = u + \frac{3u - 1}{5}.$$

Letting $v$ equal the fraction and solving for $u$ gives

$$u = v + \frac{2v + 1}{3}.$$

Letting $w$ equal the fraction and solving for $v$ gives

$$v = w + \frac{w - 1}{2}.$$

Finally, letting $z$ equal the fraction and solving for $w$ gives

$$w = 2z + 1.$$

Substituting in reverse order gives $v = 3z + 1$, $u = 5z + 2$, $t = 8z + 3$, and then values for $x$ and $y$ in terms of $z$: $y = 13z + 5$ and $x = 12 - 21z$.

Because $x$ and $y$ must be positive, $12 - 21z > 0$ and $13z + 5 > 0$ or $z < \frac{12}{21}$ and $z > -\frac{5}{13}$. The only integer value of $z$ for which this is true is $z = 0$. Thus, there is only one solution and $x = 12$ and $y = 5$. The government official bought 12 bags of red jelly beans and 5 bags of black ones.

# CHAPTER 3

# Probability and Statistics

- Odds Are...
- Many Happy Returns
- The Average Choice
- Whadaya Mean, "Mean"? ★

| + | 1 | 2 | 3 | 4 | 5 | 6 |
|---|---|---|---|---|---|---|
| 1 | 2 | 3 |   |   |   |   |
| 2 | 3 | 4 |   |   |   |   |
| 3 |   |   |   |   |   |   |
| 4 |   |   |   |   |   |   |
| 5 |   |   |   |   |   |   |
| 6 |   |   |   |   |   |   |

Black Die

White Die

First, the mathematical skills required by elementary probability problems are minimal; any eighth-grade student should know how to multiply and subtract fractions. Second, many students see this sort of problem as recreational.

Third, and perhaps most importantly, interpreting and applying probabilities is a pervasive part of out-of-school life. Many of your students are sports fans and will be interested in learning the meaning of odds and discovering something about how they are determined. All of your students will be exposed to probabilities in televised weather forecasts and the rhetoric of medical, insurance, and political advertising.

These three reasons make a good case for presenting "Odds Are..." and "Many Happy Returns" somewhere during the first-year course. This seems especially desirable for those students who will not be taking second-year algebra. You'll find these activities make good routine-breakers.

The format and technique for presenting the material in "Odds Are..." may appeal to students somewhat more than the standard text treatment. Students work their way through by constructing a table of possible outcomes for dice throws. Thus, they're able to see (and count) why the probability determination formulas work.

"Many Happy Returns" presents a problem whose solution will come as a big surprise to most students. The activity builds on the ideas and operations covered in "Odds Are...." However, because it gives a good review of "Odds Are...," a good-sized interval between using the two activities may be beneficial.

"The Average Choice" presents some elementary statistics ideas: median, mode, and mean. A teacher is depicted as citing three different "averages" as representative of the same test results, depending on whether he or she is impressing the principal or berating the class. No algebra skills are required for this activity, and for students who will not be taking additional mathematics courses, "The Average Choice" provides good tools for interpreting everyday statistical data.

The distinction between straight arithmetic means and weighted averages is essential for solving many practical problems. Unfortunately, it's a topic that is troublesome for many students. "Whadaya mean, "Mean?," uses four recognizable situations that show this distinction and how to recognize it. This is a good activity to present in conjunction with uniform motion problems.

# Odds Are...

If you throw a pair of dice twice, what's the probability of getting a 7 on the first throw and an 11 on the second throw? Take a guess: _____

The probability of an event happening is usually expressed as a fraction or percent between 0 and 1. If there's no chance of the event happening, its probability is 0. What's the probability of throwing an 8 with one die? _____

If the event is a certainty, its probability is 1. What's the probability of throwing a 1, 2, 3, 4, 5, or 6 with one die? _____ . You can calculate probabilities as follows:

$$\text{probability of an event} = \frac{\text{number of favorable outcomes}}{\text{total number of outcomes}}.$$

How many possible outcomes are there for throwing one die? _____
How many ways are there of getting a 3? _____ What is the probability of getting a 3? _____ of getting an odd number? _____ of getting a number greater than 4? _____ .

To find the probabilities of different dice throws, find all the possibilities that can occur. Suppose we have a black die and a white die. Complete the following table for all the possible sums when the two dice are thrown.

**Black Die**

| + | 1 | 2 | 3 | 4 | 5 | 6 |
|---|---|---|---|---|---|---|
| 1 | 2 | 3 |   |   |   |   |
| 2 | 3 | 4 |   |   |   |   |
| 3 |   |   |   |   |   |   |
| 4 |   |   |   |   |   |   |
| 5 |   |   |   |   |   |   |
| 6 |   |   |   |   |   |   |

(White Die)

How many possible outcomes are there? _____ How many ways can you throw a sum of 6? _____ What is the probability of throwing a sum of 6? _____ What is the probability of throwing a sum that is an

52

even number? _____

Now let's look at the probabilities for successive events. Suppose we throw the black die first and then the white die. What is the probability of getting a 2 on the black die and a 5 on the white die? The probability of a 2 on the black die is $\frac{1}{6}$ and the probability of a 5 on the white die is $\frac{1}{6}$. The probability of *both* events happening is $\frac{1}{6} \times \frac{1}{6} = \frac{1}{36}$. The table shows why this is so: Throwing the dice separately is the same as throwing both at the same time. There is only 1 possibility in 36 of getting a 2 on the black die and a 5 on the white die. We *multiply* individual probabilities to find successive probabilities.

What is the answer to the problem at the top of the previous page? _____
What's the probability of throwing "doubles" twice in a row? _____

***EXTENSION!*** Probabilities are often expressed in terms of "odds." The odds in favor of an event equal the number of favorable outcomes divided by the number of *unfavorable* outcomes. Use the probability you found previously to find the odds of getting a 7 on the first throw of two dice and an 11 on the second throw. What are the odds of getting "doubles" twice in a row?

# Teacher's Notes for Odds Are...

*The mathematical concepts in this probability topic are within easy reach of most ninth graders. In addition, because the topic is interesting to most students, this activity is a good change-of-pace activity.*

—————————————————— NCTM Standards ——————————————————

| 1 | 2 | 3 | 4 | 5 | 6 | 7 | 8 | 9 | 10 |
|---|---|---|---|---|---|---|---|---|----|
|   |   |   | • | • | • |   |   | • | •  |

### Presenting the Activity

The first question on the student page is a "teaser" that students will not be able to answer. Have them take some guesses and record the guesses on the chalkboard for later reference.

Most students will have some familiarity with probability from hearing odds quoted for sports events or the chances of rain in the weather forecast. Ask if any students know how probability and odds are found. A few may have some idea, but probably don't know the mathematics involved.

Throwing an 8 with one die is impossible, so its probability is 0. Because the six sides of a die have dots representing 1, 2, 3, 4, 5, and 6, the probability of throwing one of these numbers is 1. You may want to introduce the notation usually used for probability. For the two events,

$$P(8) = 0, \qquad P(1, 2, 3, 4, 5, \text{ or } 6) = 1.$$

Now students find probabilities that are between 0 and 1. There are six possible outcomes for throwing one die and only one of them is a 3. Thus,

$$P(3) = \frac{1}{6}.$$

There are three odd numbers, 1, 3, and 5, so

$$P(\text{odd number}) = \frac{3}{6} \quad \text{or} \quad \frac{1}{2}.$$

There are two numbers greater than 4, so

$$P(\text{number} > 4) = \frac{2}{6} \quad \text{or} \quad \frac{1}{3}.$$

These three problems should give students enough practice for finding probabilities with one die.

Students can now consider probabilities when two dice are thrown. The completed table for the possible sums of the dice is

**Black Die**

| + | 1 | 2 | 3 | 4 | 5 | 6 |
|---|---|---|---|---|---|---|
| 1 | 2 | 3 | 4 | 5 | 6 | 7 |
| 2 | 3 | 4 | 5 | 6 | 7 | 8 |
| 3 | 4 | 5 | 6 | 7 | 8 | 9 |
| 4 | 5 | 6 | 7 | 8 | 9 | 10 |
| 5 | 6 | 7 | 8 | 9 | 10 | 11 |
| 6 | 7 | 8 | 9 | 10 | 11 | 12 |

(White Die — row labels at left)

There are 36 possible outcomes and 5 ways to throw a 6. Thus,

$$P(6) = \frac{5}{36}.$$

There are 18 ways to throw an even number, so

$$P(\text{even number}) = \tfrac{18}{36} \quad \text{or} \quad \tfrac{1}{2}.$$

Give students additional practice for a pair of dice by asking them for other probabilities.

Next, students consider successive events such as the problem posed at the beginning of the activity. The probability of successive events is the *product* of the probabilities of the individual events. This will make sense to students when they consider it is less likely for two events to happen one after the other than it is for each event to happen alone. Multiplying probabilities gives a smaller probability. (See also "Pascal's Triangle.")

Now students have enough background to answer the initial question of the activity:

$$P(7) = \frac{6}{36} = \frac{1}{6}, \qquad P(11) = \frac{2}{36} = \frac{1}{18},$$

$$P(7 \text{ then } 11) = \frac{1}{6} \times \frac{1}{18} = \frac{1}{108}.$$

Thus, you are likely to throw first a 7 and then an 11 once in 108 tries. Compare this result to student's guesses at the beginning of the class.

For the probability of getting "doubles" twice in a row,

$$P(\text{doubles}) = \frac{6}{36} = \frac{1}{6},$$

$$P(\text{doubles twice}) = \frac{1}{6} \times \frac{1}{6} = \frac{1}{36}.$$

There is a much better chance (about three times as great) of throwing doubles twice than there is of throwing a 7 and then an 11. You may want to have students change some of the fractions to percentages because percentages are often used to express probabilities.

### Extension

Most gambling games and sweepstakes contests give the chances of winning in terms of odds. The Extension shows students how odds are computed and should be easy for all students.

To throw a 7 and then an 11, there is only one favorable outcome out of 108 possible outcomes. Thus, there are 107 *unfavorable* outcomes and the odds of throwing a 7 and then an 11 are 107 : 1 against. Explain that odds are written as a ratio, usually with the larger number first. Thus, these odds would be written as 107 to 1 against.

The odds of throwing doubles twice are 35 : 1 or 35 to 1 against. Have students consider other probabilities and odds and ask them to relate their experiences in dealing with probability.

# Many Happy Returns

"YOU LOSE!"

Suppose someone offered to bet that at least two people in your class have the same birthday. Would you take the bet? Whether it's a good bet for you depends on how many people are in the class. If there are more than 22 people in your class, the odds favor the person who offered the bet. And if there are 30 students in your class, the chances are better than two-to-one that at least two of you have birthdays that fall on the same date. How can we come up with this surprising number?

The probability of an event happening is usually expressed as a fraction or a decimal between 0 and 1. If there's no chance of the event happening, its probability is 0. If the event is a certainty, its probability is 1.

For the cases in between 0 and 1, we divide the number of "successes" (the events we are interested in) by the number of possible outcomes. In throwing a die there are six equally probable outcomes. So the probability of rolling a 3 with one die is _____ . The probability of rolling a number *not* 3 is _____ . Notice that the probability of success is always _____ minus the probability of failure.

We multiply to find the probability of *successive* events. What is the probability of tossing heads on the first toss of a coin *and* heads on the second toss? _____ × _____ If the coin were tossed six times, the probability of getting all heads would be

_____ × _____ × _____ × _____ × _____ × _____ .

56

The probability of *not* getting six heads in a row is

$$1 - \underline{\hspace{2cm}} \times \underline{\hspace{2cm}} \times \underline{\hspace{2cm}} \times \underline{\hspace{2cm}} \times \underline{\hspace{2cm}} \times \underline{\hspace{2cm}}.$$

Now we can get back to the birthday problem. What is the probability that the person next to you has a birthday that is different from yours? \underline{\hspace{3cm}}
Let's add a third person. What is the probability that this person's birthday will be different from yours *and* from the person next to you? \underline{\hspace{2cm}} × \underline{\hspace{2cm}}

We can continue this way until, for a class of 30 students, we find the probability of everyone having *different* birthdays, which is given by

$$\frac{364}{365} \times \frac{363}{365} \times \frac{362}{365} \times \cdots \times \underline{\hspace{1.5cm}} \times \underline{\hspace{1.5cm}} \times \underline{\hspace{1.5cm}}.$$

(You fill in the last three.) We want to find the probability of this *not* happening. That is, we want to find the probability that at least two people have the same birthday. Thus, we subtract this probability from \underline{\hspace{1cm}} and

$$P(\text{same birthday}) = \underline{\hspace{1.5cm}} - \underline{\hspace{1cm}} \times \underline{\hspace{1cm}} \times \underline{\hspace{1cm}} \times \cdots \times \underline{\hspace{1cm}} \times \underline{\hspace{1cm}} \times \underline{\hspace{1cm}}.$$

This is $1 - 0.294 = 0.706$. If you convert this to a percent, you find there is better than a 70% chance that two people in a class of 30 will have the same birthday!

***EXTENSION!*** How many students must be in a class so that you can be *certain* that at least two people will have the same birthday?

# Teacher's Notes for Many Happy Returns

*The birthday problem is always intriguing and a surprise to students because its results are so different from what "common sense" would lead us to expect. At the same time, this activity gives excellent practice in determining the probabilities of successive events.*

*Students should have worked through "Odds Are..." before attempting "Many Happy Returns." The tree diagrams in "Pascal's Triangle" are also useful as background for this activity.*

| | | | | NCTM Standards | | | | | |
|---|---|---|---|---|---|---|---|---|---|
| 1 | 2 | 3 | 4 | 5 | 6 | 7 | 8 | 9 | 10 |
| | | | • | • | • | | | • | • |

### Presenting the Activity

As mentioned previously, many students will be surprised by the probabilities cited at the beginning of the activity. Somewhat skeptical students may be further convinced of the statistical plausibility here by considering the birth and death dates of our 42 presidents. Two, Polk and Harding, were born on November 2; two, Fillmore and Taft, died on March 8; and three, Adams, Jefferson, and Monroe, died on July 4.

The activity then reviews the simplest fundamentals of probability: The probability of throwing a given number on a six-sided die is $\frac{1}{6}$. Conversely, the probability of not throwing that given number is $\frac{5}{6}$ or $1 - \frac{1}{6}$. This relationship is the key to an easy solution of the birthday problem.

Once the idea of determining single probabilities has been grasped, we can consider successive probabilities. It is here that tree diagrams of coin tosses are helpful to explain why successive probabilities are the *products* of the individual probabilities. Thus, the probability of throwing six successive heads with a true coin is

$$\frac{1}{2} \times \frac{1}{2} \times \frac{1}{2} \times \frac{1}{2} \times \frac{1}{2} \times \frac{1}{2}.$$

This, of course, is $\frac{1}{64}$. You can generalize this fundamental theorem of probability: If the probability of an event is $P_1$, and if, after it has occurred, the probability of a second event is $P_2$, the probability that both events will happen in that order is $P_1 P_2$. This principle can be extended to calculate the probability that a sequence of $n$ events will occur, given that each preceding event has occurred and that the result would be $P_1 P_2 P_3 P_4 \cdots P_n$.

Combining these two ideas allows us to solve the birthday problem. Unfortunately, successive products of the denominator 365 are very unwieldy. An ordinary deck of 52 playing cards can provide an exact analogy, yet uses numbers that are much easier to handle. Thus, you may want to have your students consider the following problem: Student number 1 picks a card (say a 9), then replaces it. Student number 2 picks a card (say a 7), then replaces it. Student number 3 picks a queen, and so on. At what point (i.e., how many students are required) is there an even chance that two people will have picked a card with the same denomination?

58

The probability for the first student is of course 1—any card will do. The probability that the second student picks a card with a different denomination is $\frac{48}{52}$ or $\frac{12}{13}$, and we know that the probability of this student picking a card with the same denomination is $\frac{1}{13}$ or $1 - \frac{12}{13}$. When we introduce the third student, his or her probability of selecting a denomination different from either of the first two students is $\frac{12}{13} \times \frac{11}{13}$. Thus, the probability of the third students picking a card of the same denomination is $1 - \frac{12}{13} \times \frac{11}{13}$. After only five students, the probability of two people picking cards with the same denomination is

$$1 - \frac{12}{13} \times \frac{11}{13} \times \frac{10}{13} \times \frac{9}{13} = 1 - 0.416 = 0.584.$$

This is better than an even chance.

If our criterion were picking the same suit instead of the same denomination, there would be a much better than even chance after the third student and an absolute certainty at five students.

Applying this same reasoning to the birthday problem, we have, for a class of 30 students,

$$P(\text{two have same birthday}) = 1 - \frac{364}{365} \times \frac{363}{365} \times \cdots \times \frac{336}{365}$$
$$= 1 - 0.294$$
$$= 0.706.$$

As shown by the accompanying graph, the probability goes to 80% when the class size becomes 35 and nearly 90% when the class size becomes 40.

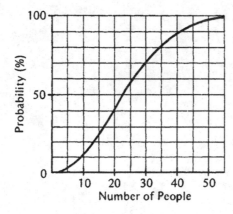

### Extension

The question posed in the Extension is not entirely without guile. It requires no knowledge of probability operations—just common sense. The curve shown in the preceding graph is *not* asymptotic; it reaches 100% at 366 or 367, depending on whether leap years are included.

# The Average Choice

An algebra teacher told the head of the math department that his class had done very well on the final exam—the average score was 92. The teacher then told the class they would have to work much harder the next term—the average score on their final exam was only 79. The teacher also told one student that her score of 85 on the final exam was the average score.

What's going on here? How can the same test have an average score of 92, 79, and 85? Let's look at the scores and see how these averages were found. The scores were

92  89  86  85  92  86  74  80  92
70  60  94  65  80  41  72  65  89
59  72  92  89  70  89  92

Begin by organizing the scores in the accompanying chart at the right. Use tally marks ( *HH* ) to record how many there are of each score. This results in what is called a *frequency distribution*.

| Score | Frequency |
|-------|-----------|
| 94    |           |
| 92    |           |
| 89    |           |
| 86    |           |
| 85    |           |
| 80    |           |
| 74    |           |
| 72    |           |
| 70    |           |
| 65    |           |
| 60    |           |
| 59    |           |
| 41    |           |

Now we can use the frequency distribution to help find averages. Which score occurs most often? _____ This score is called the *mode*. Which score is in the middle? That is, which score has the same number of values greater than it as there are less than it? _____ This score is the *median*.

The *mean* is a third type of average. It is found by adding all the scores and then dividing by the number of scores. What is the mean for the scores? _____

_____ The mean, median, and mode are all *measures of central tendency*. They are all single numbers, used to give us an idea of what a set of numbers is like. If the mean, median, and mode all have nearly the same value, the set of numbers is usually well balanced about its central value.

Sometimes the numbers in a frequency distribution are grouped into *class intervals* and then graphed. One such graph is called a *histogram*. The horizontal axis shows the score intervals and the vertical axis shows the frequency. Complete the following table and histogram. (Use the scores listed in the preceding text.)

| Class Interval | Frequency |
|---|---|
| 91–100 | 6 |
| 81–90 | |
| 71–80 | |
| 61–70 | |
| 51–60 | |
| 41–50 | |

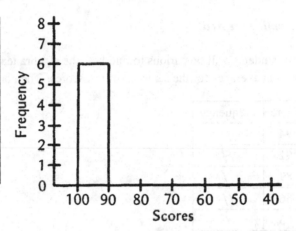

*EXTENSION!* The test scores on an English exam were 84, 78, 98, 89, 80, 73, 91, 87, 54, 73, 84, 95, 73, 89, 91, 87, 73, 62, 89, 84, 75, 83, 84, 75, 73, 67, 91, 73, 62, and 83. Make a frequency distribution and find the three measures of central tendency. Then group the scores using the class intervals shown in the table and draw a histogram.

# Teacher's Notes for The Average Choice

*Means, medians, and modes are often misunderstood and misused. Yet, as this activity shows, a well-spent half hour can clear up the topic. "The Average Choice" also introduces other statistical techniques and provides a good change of pace from solving equations. No algebraic background is required.*

―――――――――――――――――― NCTM Standards ――――――――――――――――――

| 1 | 2 | 3 | 4 | 5 | 6 | 7 | 8 | 9 | 10 |
|---|---|---|---|---|---|---|---|---|---|
|   |   |   | • | • | • |   |   | • | • |

### Presenting the Activity

Most students will be curious to find how the algebra teacher managed to give three very different averages for the same set of test scores. The completed frequency distribution is

| Score | Frequency |
|-------|-----------|
| 94 | / |
| 92 | ℍℍ |
| 89 | //// |
| 86 | // |
| 85 | / |
| 80 | // |
| 74 | / |
| 72 | // |
| 70 | // |
| 65 | // |
| 60 | / |
| 59 | / |
| 41 | / |

Students simply go through the list of scores on the student page and put tally marks after the appropriate numbers in the table.

That the mode is 92 is immediately obvious from the frequency distribution. The median is also easy to find: Students can count to find there are 25 scores. Thus, there will be 12 scores above the median and 12 scores below it. The median is 85. The mode and median are *positional* averages. They are determined by their positions in the distribution rather than by their numerical values.

When most people use the word "average," they are thinking of the mean. Students now realize that the mean is only one type of average. The mean is an *arithmetic* average—it is determined by the numerical values in the frequency distribution. The sum of the scores is 1975. Dividing this by 25, students find the mean is 79. Thus, students have found the three averages used by the algebra teacher. Discuss how the teacher used the three averages to mislead the people he was talking to. A good example is shown in

Darrel Huff's *How to Lie with Statistics*. The president of a company claims the average salary in his firm is $5700, but notice the frequency distribution!

|        |        |                |
|--------|--------|----------------|
|        | 45,000 | /              |
|        | 15,000 | /              |
|        | 10,000 | //             |
| Mean   | 5,700  | /              |
|        | 5,000  | ///            |
|        | 3,700  | ////           |
| Median | 3,000  | /              |
| Mode   | 2,000  | ⊮⊬ ⊮⊬ //       |

During the 1981 baseball strike, the audience of a CBS newscast heard a management spokesman (using the mean) say the average player was paid approximately $130,000 per year. A player representative (using the median), stated that the players averaged $105,000.

Other information can be found from the frequency distribution. For example, if all the scores over 90 are As, students can find what percent of the class scored an A on the exam: $\frac{6}{25} = \frac{24}{100} = 24\%$. Alternatively, they can find what percent scored higher than the mean: $\frac{15}{25} = \frac{60}{100} = 60\%$. Thus, the mean is influenced by very large or very small values. This helps explain why three measures of central tendency give a more accurate picture of the set of numbers. Give an example of a set of numbers well balanced about its central value and have students find the mean, median, and mode.

By grouping the numbers into class intervals and drawing a histogram, students are able to see the list of scores:

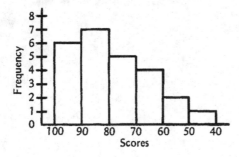

Point out that organizing values into class intervals is particularly helpful when there are a large number of values.

*Extension*

All students should be able to complete the Extension. The frequency distribution is

| Score | Frequency |
|-------|-----------|
| 98 | / |
| 95 | / |
| 91 | /// |
| 89 | /// |
| 87 | // |
| 84 | //// |
| 83 | // |
| 80 | / |
| 78 | / |
| 75 | // |
| 73 | ⊥⊥⊥ / |
| 67 | / |
| 62 | // |
| 54 | / |

The mode is 73, the median is 83, and the mean is $2400 \div 30 = 80$.

The histogram is

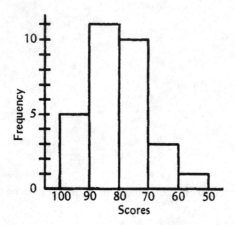

Students can compare this histogram with the preceding one.

You may want to have students prepare a frequency distribution from actual data for your class, such as test scores, attendance, or heights of class members.

# Whadaya Mean, "Mean"?

Last Sunday Carl took his little brother to camp. He was glad to do it; the kid would be gone for 6 weeks. Traffic was light on the way to the camp, and Carl was pleased to find that the 50-mi trip had taken exactly 1 hr. What was his average speed for the trip to camp? _____

Traffic was much heavier on the way home. That trip took 2 hr. What was his average speed for the trip home? _____

What was his average speed for the round trip; that is, driving from home to camp and back home? _____

Did you answer 37.5 mph? A lot of people do. Stop and think a minute. How long did he drive while averaging 50 mph? _____ How long did he drive while averaging 25 mph? _____ Should the two averages be given "equal weight?" _____

Let's look at the problem this way: What is the distance for the round trip? ___ _____ What was Carl's total driving time for the round trip? _____

Now what was his average speed for the round trip? _____

What you have done here is assign the proper weights to the terms when you're trying to find the right "average."

Look at it this way: Suppose on four exams this semester you scored 90 on three of them and 60 on the other one. You wouldn't be very pleased if your teacher computed your average by adding 90 to 60 and dividing by 2. What "average score" would you get using this method? _____

How would you compute your average grade? What does it come out to? ____

One more example should give you a good idea of how weighted averages work and how useful they are. Wild Willie is all set to write the great American novel. Knowing that first things should come first, he goes out and buys $6 worth of pencils—$2 worth of pencils that cost 5¢ each, $2 worth of pencils that cost 10¢ each, and $2 worth of pencils that are $12\frac{1}{2}$¢ each. What is the average price per pencil? _____

_____

_____

*EXTENSION!* In the first problem, where Carl takes his brother to camp, the distance from home to the camp is 50 mi. But it could have just as easily been 60 or 100 mi. *As long as the distance on both legs of the trip remains the same*, the method works for whatever rates you assign for each part of the round trip. Show why this is so.

# Teacher's Notes for Whadaya Mean, "Mean"?

*"The Average Choice" introduces the ideas of means, medians, and modes in a way that depends on no algebraic skills. It is an ideal predecessor to this activity which does require some equation-solving ability.*

*The first and last problems are specific applications of the harmonic mean. The middle problems are more general situations that require simple weighted averages. Of greatest importance is that students are clearly shown why harmonic means or weighted averages are needed when straight arithmetic means give distorted results.*

--- NCTM Standards ---

| 1 | 2 | 3 | 4 | 5 | 6 | 7 | 8 | 9 | 10 |
|---|---|---|---|---|---|---|---|---|----|
|   |   |   | • | • | • |   |   | • | •  |

### Presenting the Activity

Many people will quickly answer the first question with 37.5 mph as the average speed for the round trip. Their reasoning is this: 50 mph on the way out; 25 mph on the way in; halfway between them is 37.5 mph. However, did Carl drive at these rates for equal amounts of time? No, the 25-mph rate has twice the influence on the round-trip average rate as does the 50-mph rate. This is borne out by using $D = RT$ for the round trip. We get

$$50\,\text{mi} + 50\,\text{mi} = R(1\,\text{hr} + 2\,\text{hr}) \quad \text{and} \quad R = 33\frac{1}{3}\,\text{mph}.$$

Some students will accept this with something of a "Well, if you say so" feeling. Naturally, they're not entirely convinced nor at all clear yet as to why it works this way. The next example, however, strikes a little closer to where they live and they have a vested interest in seeing which of the two procedures yields the more accurate results. Adding 90 to 60 and dividing by 2 gives an "average score" of 75. Giving each test equal weight produces

$$(90 + 90 + 90 + 60) \div 4 = 82.5,$$

a significantly better grade. Have students take their time and think about this problem for a minute or so. The problem is a straightforward application of a weighted average.

Now present this one: A teacher gives four quizzes and one final exam. The quizzes each count for 15% in determining final grades, and the final exam counts 40%. If a student's quiz scores are 80, 75, 85, and 80, and 87 is the final exam score, the final grade is

$$\frac{15(80) + 15(75) + 15(85) + 15(80) + 40(87)}{100}.$$

This equals 82.8%. Note that the quizzes average exactly 80%. The final exam counts for 40% of the grade and was 7% higher than the quiz grade. 0.4 times 7% shows the 2.8% increase in the final grade.

Students should now feel comfortable enough so they can easily solve the next problem in the following way: At 5¢ each, Willie buys 40 pencils for $2, 20 pencils at 10¢ each ($2), and 16 pencils at $12\frac{1}{2}$¢ each ($2). He has thus bought 76 pencils for $6, so the average cost is 7.89¢ per pencil. Point out that had you simply taken the average of 5¢, 10¢, and $12\frac{1}{2}$¢, you would have arrived at an incorrect figure of 9.17¢.

### Extension

The Extension to this activity is an especially flexible one. Have students consider the problem in general terms. For the first leg of the trip the time

$$T_1 = \frac{D}{R_1}.$$

For the trip back home, the time

$$T_2 = \frac{D}{R_2}.$$

The time for the total trip then is

$$
\begin{aligned}
T &= T_1 + T_2 \\
&= \frac{D}{R_1} + \frac{D}{R_2} \\
&= \frac{D(R_1 + R_2)}{R_1 R_2}.
\end{aligned}
$$

However, the rate for the entire trip is

$$
\begin{aligned}
R &= \frac{2D}{T} \\
&= \frac{2D}{\frac{D(R_1 + R_2)}{R_1 R_2}} \\
&= \frac{2R_1 R_2}{R_1 + R_2}.
\end{aligned}
$$

The $D$s have cancelled out and $R = \frac{2R_1 R_2}{R_1 + R_2}$.

This is the reciprocal of the average of the reciprocals of $R_1$ and $R_2$. Such an average is called the *harmonic mean*. A progression of numbers is said to be harmonic if any three consecutive members of the progression, $a$, $b$, and $c$ have the property that

$$\frac{a}{c} = \frac{a - b}{b - c} \quad \text{or} \quad a(b - c) = c(a - b).$$

Dividing by $abc$, we get $\frac{1}{c} - \frac{1}{b} = \frac{1}{b} - \frac{1}{a}$.

This relationship shows that the reciprocals of a harmonic progression are in an arithmetic progression, as with $\frac{1}{a}$, $\frac{1}{b}$, and $\frac{1}{c}$. When three terms are in an arithmetic progression, the middle term is their mean. Thus, $\frac{1}{b}$ is the arithmetic mean between $\frac{1}{a}$ and $\frac{1}{c}$; $b$ is the harmonic mean between $a$ and $c$.

Expressing $b$ in terms of $a$ and $c$,

$$\frac{2}{b} = \frac{1}{a} + \frac{1}{c} \quad \text{and} \quad b = \frac{2ac}{a+c}.$$

Compare this with $\frac{2R_1R_2}{R_1+R_2}$.

# CHAPTER 4

# Problem Solving

- More Posers
- Your Number, Please
- Optimizing a Paper Route
- Angles on a Clock
- Digit Problems Revisited ★

As mentioned in the general introduction, problem solving is the heart of these activities. I am, therefore, much more concerned with students forming a strategy—an ability to read and plan an attack—than I are concerned with individual algebraic manipulation skills. Accordingly, the first three activities in this section require little or no algebra preparation, but they do require careful reading.

Mathematically, "More Posers" requires only the simplest arithmetic skills and some common sense. If the activity is given late in the course, some students may become bogged down with calculations that aren't at all necessary—particularly in the last two probability problems. You may wish to use "Posers" in *Making Pre-Algebra Come Alive* as either a warm-up or supplement.

The following two activities, "Your Number, Please" and "Optimizing a Paper Route," are very unlike most textbook problems. Some ingenuity is required, but a sound plan and persistent followup are more important.

"Your Number, Please" emphasizes deriving a general formula from given data. "Optimizing a Paper Route" provides practice in locating ordered pairs of numbers on a coordinate grid. However, these two mathematical concepts are less important than the development of mathematical modeling skills. Very few students are able to set up a mathematical model of a physical situation, and yet this is one of the most important skills in solving and analyzing problems.

"Angles on a Clock" has the properties of an incipient classic. What at first appears as a problem that must be very difficult to calculate with precision, turns out to be an application of our old friend: distance = rate × time. Moreover, because this is circular motion, there is a "correction factor" that can be uniformly inserted—something that can't be done with the typical train *A* meets or overtakes train *B* sort of problem.

The last activity in this section, "Digit Problems Revisited," will be intriguing to your better students, but some care will be necessary to keep from frustrating your less capable students. The actual equations that students must solve are quite straightforward—indeed ordinary—but translating the stated problems into these simple equations is a challenge that will separate the sheep from the goats. Like most challenges, successfully cracking these nuts will give real satisfaction to students.

# More Posers

A cargo ship ties up to a pier at Seattle's waterfront. It is 20 ft from the ship's deck to the surface of the water. A rope ladder hangs over the side of the ship. The rungs of the ladder are 1 ft apart and the water just covers the bottom rung of the ladder. During the night the tide comes in and the water level at the side of the pier rises 8 ft. How many rungs of the ladder are covered in the morning?

_____

A spider at the bottom of a 50-ft well strikes out to discover life at the top. Each day the spider climbs 5 ft, but then each night it slips back 3 ft. How long will it take the spider to get out of the well? _____

A 1-mi-long freight train traveling 30 mph enters a tunnel at 12 noon. The tunnel is 2 mi long. At what time is the train clear of the tunnel? _____

Stranded in a life raft, a man has a canteen of water and a 16-oz bottle of vegetable juice. To conserve his "food," he drinks one ounce of the vegetable juice the first day, and fills the bottle by pouring 1 oz of water into it. The second day he drinks 2 oz of the water/juice mixture and replaces it with 2 oz of water. The third day it's 3 oz and so on. By the time he empties the bottle, how much water has he drunk from it? _____

One afternoon Lisa received four utility bills in the mail. She immediately sat down and wrote four checks—to the electric company, water department, and so forth. Before she could put the checks into the proper return envelopes, the telephone rang. While she was on the phone her mischievous little boy, David, put a check into each envelope and sealed it. What is the probability that David put exactly three checks into the correct envelopes? (No, David can't read.) ___

***EXTENSION!*** Alfred was convicted of a crime in the small, island republic of Cuny. In that country the convicts determine their sentences in the following way: Alfred is given two identical jars and 100 small balls—50 white and 50 black. He then puts the balls into the jars—as may of each color into each jar as he wishes. The only restriction is that all 100 balls have to somehow be put into the two jars. Alfred is then blindfolded and told to draw one ball from one of the jars. If he picks a white ball he goes free; if he picks a black ball he is executed. How can he arrange the balls to give himself the best chance of going free? (The balls inside the jars will be well stirred before he gets to pick and he can't tell which jar is which.)

# Teacher's Notes for More Posers

*As stated in the general introduction to this volume, problem solving cannot be overemphasized in your curriculum and teaching strategies. This is one of only a few areas where math educators are in nearly unanimous agreement. Mathematics purists may wish it otherwise, but the great majority of students study math primarily to learn how to solve problems.*

*This activity should be presented first in the problem-solving section and, if students haven't already been exposed to "Posers" from **Making Pre-Algebra Come Alive**, the two activities complement each other nicely and are springboards for lively, enjoyable discussions.*

―――――――――――― NCTM Standards ――――――――――――

| 1 | 2 | 3 | 4 | 5 | 6 | 7 | 8 | 9 | 10 |
|---|---|---|---|---|---|---|---|---|----|
|   |   |   |   |   | • | • | • | • | •  |

## *Presenting the Activity*

The first problem, as is the case with the first problems in "Posers," requires no mathematical skill at all. What is required, of course, is an ability to read and, more important, the patience to think a minute (or less) about what you've just read. Whether the incoming tide raised the water level 4 ft, 8 ft, or 16 ft is beside the point. The ship is floating on the water and so the water still covers only one rung of the ladder. The point to be made then is that mathematical skills may be ancillary to problem-solving skills; the ability to count, add, multiply, and find second derivatives, standard deviations, and matrix inverses is useful only when you know what and why you're counting, adding, ... .

The second problem is only a step removed from the first one. Students will quickly see that the spider's net progress per day is $5 - 3 = 2$ feet, and that dividing 50 feet by 2 feet per day quickly yields 25 days. Ah, but look again! On the morning of the twenty-fourth day Mr. Spider has advanced 46 feet. His five-foot advance this day will take him over the top and he won't slide back to 48 feet.

The third problem goes one more step. No trickiness here, but some analysis is required, and it again takes precedence over the elementary algebra skill required. The train is 1 mi long and the tunnel is 2 mi long. Therefore, the front of the train must travel 3 mi before its rear end clears the tunnel. Given a steady 2 min per mile, this will occur at 12:06 p.m.

The life-raft problem is also not a trick, but it does afford an opportunity for students to make things unnecessarily difficult for themselves. Students can easily and understandably become bogged down in calculating how much water the man drinks on day $n$, $n + 1$, etc. However, all a person needs to know is that he adds one ounce of water to the bottle the first day, two ounces the second, ..., fifteen ounces the fifteenth day. And throughout the ordeal he drinks it all—which is

$$1\,oz + 2\,oz + \cdots + 15\,oz.$$

This is simply

$$\frac{15}{2}(1 + 15) = 120\,oz.$$

Students who find this shortcut intriguing will enjoy the story of Gauss in "Odd-Order Magic Squares" in *Making Pre-Algebra Come Alive*.

The final two problems in this activity deal with probability—though again, common sense takes precedence over mathematical operations. In the problem that matches checks with envelopes, the probability of getting exactly three correct matches is zero. If three of the checks and envelopes are correctly paired, then so must be the fourth!

### Extension

If Alfred distributes the balls evenly between the two jars, he obviously has an even chance of drawing a white ball. But let's see what happens if he places a single white ball in the first jar and the remainder of the balls in the second jar. His chances of going free are very nearly 75%. He has a 50/50 chance of selecting jar number one. This contains only a white ball, and so he goes free. If he selects jar number two, he still has nearly an even chance—49/99 to be exact. With this arrangement his mathematical probability of picking a white ball is

$$\frac{1}{2} \times 1 + \frac{1}{2} \times \frac{49}{59} = \frac{74}{99}.$$

If Alfred can somehow talk the authorities into letting him use three or four jars, his chances of going free become even better. With three jars, the first two each containing a single white ball, the probability is given by

$$\frac{1}{3} \times 1 + \frac{1}{3} \times 1 + \frac{1}{3} \times \frac{48}{98} = 0.83.$$

With four jars, we have

$$\frac{1}{4} \times 1 + \frac{1}{4} \times 1 + \frac{1}{4} \times 1 + \frac{1}{4} \times \frac{47}{97} = 0.871,$$

very nearly $\frac{7}{8}$.

# Your Number, Please

Modern telephone networks are very complex. It's now possible to place a telephone call to almost anywhere in the world. However, the first telephone networks were much simpler. A telephone call was possible only if both telephones were directly connected by a line. Thus, six lines were needed to connect four phones as shown:

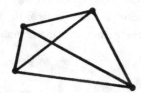

Complete the following table that shows how many lines would be needed for different numbers of telephones. Try to find a pattern.

| Telephones | 2 | 3 | 4 | 5 | 6 | 7 | 8 | 9 | 10 |
|---|---|---|---|---|---|---|---|---|---|
| Number of Lines | | | | | | | | | |

Write a formula for the number of lines needed for $n$ telephones: _____

Soon, switching centers were developed. Each telephone was connected to the switching center and an operator connected lines to complete a call. The following figure shows where a switching center $Y$ could be located for the four telephones shown previously.

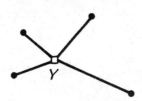

Is more or less wire needed to complete a single call when a switching center is used? _____ For which network would be the *total* amount of wire be less? _____

The next figure shows a scale map of the locations of nine telephones.

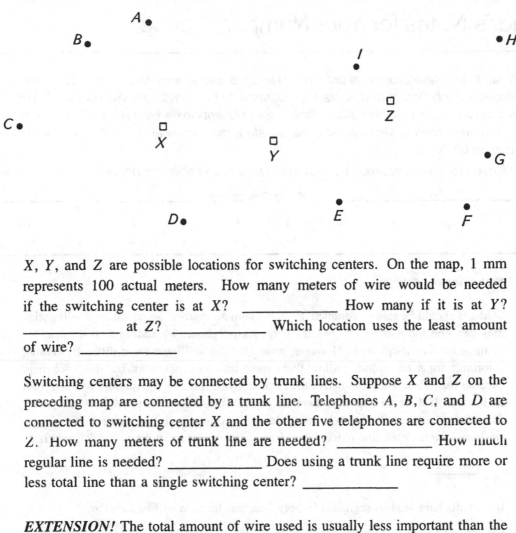

$X$, $Y$, and $Z$ are possible locations for switching centers. On the map, 1 mm represents 100 actual meters. How many meters of wire would be needed if the switching center is at $X$? _____ How many if it is at $Y$? _____ at $Z$? _____ Which location uses the least amount of wire? _____

Switching centers may be connected by trunk lines. Suppose $X$ and $Z$ on the preceding map are connected by a trunk line. Telephones $A$, $B$, $C$, and $D$ are connected to switching center $X$ and the other five telephones are connected to $Z$. How many meters of trunk line are needed? _____ How much regular line is needed? _____ Does using a trunk line require more or less total line than a single switching center? _____

***EXTENSION!*** The total amount of wire used is usually less important than the total cost of the network. Suppose telephones cost $15 each, switching centers cost $60 each, regular line is 60¢ per meter, and trunk line is $1.25 per meter. Find the cost of the previously discussed network. Then locate one or more switching centers to produce a less expensive network.

# Teacher's Notes for Your Number, Please

*This activity reinforces several areas of pre-algebra and provides an excellent change of pace from equation solving. Although the concepts and operations required by the activity are elementary, it focuses on two extremely important problem-solving skills. First, it asks students to derive a general formula from empirical data; second, it has students visualize and manipulate a mathematical model that represents a physical and economic problem.*

*Students will need to have rulers measured in millimeters to complete the activity.*

―――――――――――――――――― NCTM Standards ――――――――――――――――――

| 1 | 2 | 3 | 4 | 5 | 6 | 7 | 8 | 9 | 10 |
|---|---|---|---|---|---|---|---|---|----|
|   | • | • |   |   | • | • | • | • |    |

### Presenting the Activity

Students should be able to complete the table with no trouble. Some students will notice they can find the number of lines for (say) six telephones by adding 5 to the number of lines for five telephones. However, most students will have some difficulty finding a formula for *n* telephones. Allow them some time to experiment, but don't let them become frustrated. They can derive the formula by considering the number of lines from each point. If there are *n* points, there are $(n-1)$ lines from each point. It would seem, then, that there would be $n(n-1)$ lines. However, because each line connects two points, we have counted each line twice. Therefore, we must divide by 2 and the formula is

$$\frac{n(n-1)}{2}.$$

If students have studied segments in polygons, this formula will be familiar.

By comparing the two figures for four telephones, students should see that more wire is needed to place a single call when a switching center is used. However, when a switching center is used, the *total* amount of wire for the network is less.

The remainder of the activity considers a network of nine telephones. Students begin by considering three possible locations for switching centers. Students must measure the amount of wire in millimeters for each possible location. Then they must change the number of millimeters to meters. The switching center at $X$ would require 48,000 m of wire; the one at $Y$ would require 42,400 m; the one at $Z$, 44,300 m. Thus, locating the switching center at $Y$ would use the least amount of wire.

When trunk lines are introduced, the total amount of wire is reduced. For the trunk line between $X$ and $Z$, as described on the second student page, 6000 m of trunk line and 25,600 m of regular line are needed. This is a total of only 31,600 m of wire. Thus, if the total amount of wire is the major concern, using trunk line provides the more efficient network.

### Extension

The Extension continues the activity of the advantage of using switching centers by considering the costs of materials for the various networks discussed. With a single switching center at $Y$ (the location using the least amount of wire), the cost of materials is $135 for telephones, $60 for the switching center, and $25,440 for the wire. This is a total cost of $25,635.

For the network with a trunk line between $X$ and $Y$, there is $135 for telephones, $120 for two switching centers, $15,360 for regular wire, and $9300 for trunk line. Thus, the total cost for this network is $24,915.

This Extension is very open-ended and your discussion can proceed along several lines. For example, you may want to discuss the costs in more detail by considering a 10¢ decrease in the price of regular wire or a 25¢ increase in the price of trunk line. You could discuss how adding another telephone would change the costs of the two networks, depending on where the new phone was located. Alternatively, as indicated on the student pages, students can experiment with trying to locate one or more switching center to reduce the cost of materials.

# Optimizing a Paper Route

Jenny has 10 customers on her Sunday morning paper route. The location of each customer is shown on the following graph

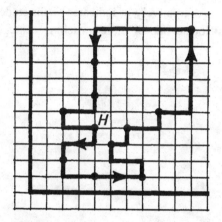

Her home is at $H(4, 4)$. The grid lines represent streets and the heavy line shows the route Jenny walks each Sunday. How many blocks does she walk? _____

Jenny carries all 10 of her papers when she leaves her house and each paper is very heavy. She wonders if she has planned her route the easiest way. Find a shorter route than the one Jenny walks. Draw your route on the following grid.

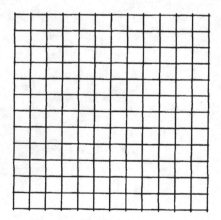

How many blocks long is your route? _____

Jenny decides the shortest route may not be the easiest. For example, her first stop is at (2, 2), so she has to carry all 10 papers for four blocks. If her first stop were at (5, 3), she would only have to carry the 10 papers for two blocks. To find the work load for her present route, she makes up the chart

| For Each Block Walked | Points |
|---|---|
| with no papers | 1 |
| with 1 or 2 papers | 2 |
| with 3 or 4 papers | 3 |
| with 5 or 6 papers | 4 |
| with 7 or 8 papers | 5 |
| with 9 or 10 papers | 6 |

The route with the lowest total points should be the easiest route. Complete the following tables for Jenny's present route and the route you found.

Jenny's Route

| From (4, 4) | Blocks | Papers | Points |
|---|---|---|---|
| to (2, 2) | | | |
| to (4, 1) | | | |
| to (7, 1) | | | |
| to (5, 3) | | | |
| to (6, 4) | | | |
| to (8, 5) | | | |
| to (10, 10) | | | |
| to (4, 8) | | | |
| to (4, 6) | | | |
| to (2, 5) | | | |
| to (4, 4) | | | |

Your Route

| From (4, 4) | Blocks | Papers | Points |
|---|---|---|---|
| | | | |
| | | | |
| | | | |
| | | | |
| | | | |
| | | | |
| | | | |
| | | | |
| | | | |
| | | | |
| to (4, 4) | | | |

81

How many work-load points are there for Jenny's route? _____

How many are there for your route? _____

One Sunday Jenny splits her route. She takes only five papers and delivers them to (6, 4), (5, 3), (7, 1), (4, 1), and (2, 2) and then returns home. Then she delivers the other five papers to (2, 5), (4, 6), (4, 8), (10, 10), and (8, 5) and then returns home. How many blocks does she walk for this route? _____

What are her total work-load points this way? _____

*EXTENSION!* Find a shorter split route that takes fewer work-load points.

# Teacher's Notes for Optimizing a Paper Route

*Most students need practice in locating ordered pairs when first introduced to coordinate graphing. This activity gives that practice and, more importantly, helps develop students' problem-solving skills. It may also be one of their first experiences in analyzing a practical problem by setting up a mathematical model.*

*The only prerequisite is an introduction to locating ordered pairs on a graph. Spare graph paper is useful to have on hand.*

| | | | | NCTM Standards | | | | | |
|---|---|---|---|---|---|---|---|---|---|
| 1 | 2 | 3 | 4 | 5 | 6 | 7 | 8 | 9 | 10 |
| • | • | • | | | • | | | | |

### Presenting the Activity

Briefly discuss how the coordinate grid represents streets and how ordered pairs indicate the locations of Jenny's customers. Jenny's route is 42 blocks long.

Students will need some time to experiment with different routes. Most students will find shorter routes strictly by trial and error. A few of your better students may realize that the way to find a shorter route is to go directly from (2, 5) to (2, 2). Two shorter routes, each 40 blocks long, are

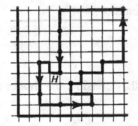 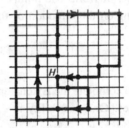

Work through the first few entries in the work-load table for Jenny's route to be sure students understand how to complete it. A completed table is

| From (4, 4) | Blocks | Papers | Points |
|---|---|---|---|
| to (2, 2) | 4 | 10 | 24 |
| to (4, 1) | 3 | 9 | 18 |
| to (7, 1) | 3 | 8 | 15 |
| to (5, 3) | 4 | 7 | 20 |
| to (6, 4) | 2 | 6 | 8 |
| to (8, 5) | 3 | 5 | 12 |
| to (10, 10) | 7 | 4 | 21 |
| to (4, 8) | 8 | 3 | 24 |
| to (4, 6) | 2 | 2 | 4 |
| to (2, 5) | 3 | 1 | 6 |
| to (4, 4) | 3 | 0 | 3 |

The total number of work-load points for Jenny's route is 155. For the preceding two shorter routes, the total number of work-load points in each is 142.

Note that the number of work-load points will change if the routes are traveled in the opposite direction. For Jenny's route, the work-load points increase to 160 if the route goes from (2, 5) to (4, 6) to (4, 8), and so on. For the two shorter routes, the work-load points for the one at the left increase to 155 if the route is traveled in the opposite direction; for the route at the right, the increase is to 159. Thus, it's possible to have a shorter route than the original one with a greater number of work-load points.

Have students compare their routes to find the route with the smallest number of work-load points.

The split route presented on the student page is

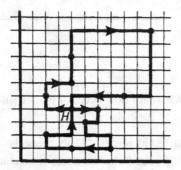

The route is 46 blocks long and has 104 work-load points. Again, the direction of each split route influences the number of work-load points. It's possible to have 110, 120, and 126 work-load points depending on the direction of the routes.

### Extension

Most students should be able to complete the Extension if they consider the shorter routes they found in the previous cases. One possibility is

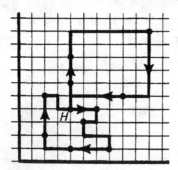

In this case, the route is split into six papers and four papers. The route is 44 blocks long and has 99 work-load points.

You may want to have students make up a different table for work-load points, assigning points for each paper carried rather than for each two papers. They can then investigate different split routes using their new table.

# Angles on a Clock

In this exercise you will learn a method with which you can easily calculate the exact times when the hands of a clock form certain angles. Let's start with an angle of zero degrees; in other words, when the minute hand is just passing the hour hand. You probably already know one such time when the minute and hour hands are exactly lined up. The minute hand passes the hour hand at _____ (what time?).

The other times when the minute hand passes the hour hand are a little trickier to figure. They're not exactly on the hour. For that matter, they're not exact minutes. At what time after 1:00 does the minute hand first pass the hour hand? A close first guess would be _____ .

At that time the minute hand has traveled to where the hour hand was at 1:00. But the hour hand hasn't been standing still, has it? While the minute hand moved from the 12 to the 1, the hour hand moved forward a little, so at 1:05 the minute hand hasn't quite caught up yet.

How can we figure exactly when the minute hand passes the hour hand? We can treat it as a difference-in-rate problem—just as we could calculate when a fast-moving train will overtake a slower-moving train that has a head start. To do this we use the formula $d = rt$ (distance equals rate times time). If we say that the hour hand moves at a rate $r$, how fast does the minute hand move? ___

Let's call the minutes on a clock *markers* and call $d$ the number of markers the minute hand must pass before it catches up to the hour hand. Remembering that at 1:00 the hour hand has a 5-min (marker) head start on the minute hand, how many markers will the hour hand pass before the minute hand catches up to it?

---

Solving $d = rt$ for $t$, we get $t = \frac{d}{r}$. For the hour hand, $t =$ _____ . For the minute hand, $t =$ _____ .

Now we know that the time that it will take for the minute hand to catch up to the hour hand is the same for both hands. Thus, $t$ for the minute hand is the same as $t$ for the hour hand, so we see that the two equations can now be combined as _____ .

Solving for $d$, we find $d =$ _____ . Thus, the minute hand passes the hour hand at _____ .

Now find when the minute hand first passes the hour hand after 7:00. In this case, if $d$ is the distance the minute hand travels, the hour hand travels _____ .

Because the two $t$s are equal, $\frac{d}{12r} =$ _____ . Thus, the minute hand passes the hour hand at _____ .

***EXTENSION!*** At what time after 10:00 do the hands form a 180° angle (i.e., form a straight line)? In addition to 3:00 and 9:00, when do the hands form right angles?

# Teacher's Notes for Angles on a Clock

*This activity is a good test of students' problem-solving abilities. The solution to the problem posed is quite straightforward once students recognize that the circular motion of clock hands can be viewed as a traditional linear motion problem.*

---------- NCTM Standards ----------

| 1 | 2 | 3 | 4 | 5 | 6 | 7 | 8 | 9 | 10 |
|---|---|---|---|---|---|---|---|---|----|
| • | • | • |   |   |   |   |   |   |    |

### Presenting the Activity

Ask students at what time (exactly) will the hands of a clock overlap after 1 o'clock. Your students' first approximation is simply 1:05. When you remind them that the hour hand moves uniformly, they will begin to estimate the answer to be between 1:05 and 1:06. They will realize that the hour hand moves through an interval between minute markers every 12 min. Therefore it will leave the interval 1:05–1:06 at 1:12. This however doesn't answer the original question about the exact time of this overlap.

The best way for students to understand the movement of the hands of a clock is to consider that the hands travel independently around the clock at uniform speeds. The minute markings on the clock (markers) serve to denote distance as well as time. An analogy should be drawn here to the uniform motion of trains—a popular and overused topic for verbal problems in an elementary algebra course. A problem involving a fast train overtaking a slower one is appropriate. The analogy should be drawn between specific cases rather than mere generalizations. It might be helpful to have the class find the distance necessary for a train traveling at 60 mph to overtake a train that has a head start of 5 mi and is traveling at 5 mph.

Have the class consider 1 o'clock as the initial time on the clock. Our problem is to determine exactly when the minute hand overtakes the hour hand after 1 o'clock. Consider the speed of the hour hand to be $r$. Then the speed of the minute hand must be $12r$. We seek the distance, measured by the number of markers traveled, that the minute hand must travel to overtake the hour hand.

Let us refer to this distance as $d$ markers. Hence, the distance that the hour hand travels is $d - 5$ markers, since it has a 5-marker head start on the minute hand. If $d = rt$, the times required for the minute hand, $\frac{d}{12r}$, and for the hour hand, $\frac{d-5}{r}$, are the same. Therefore,

$$\frac{d}{12r} = \frac{d-5}{r}.$$

Multiply both sides by $12r$ to get

$$d = 12d - 60,$$
$$11d = 60,$$
$$d = \frac{12}{11} \cdot 5$$
$$= 5\frac{5}{11}.$$

Thus, the minute hand will overtake the hour hand at exactly $1:05\frac{5}{11}$.

After students have discovered that the overlap between 7:00 and 8:00 is $7:38\frac{2}{11}$, consider the expressions $d = \frac{12}{11} \cdot 5$ and $d = \frac{12}{11} \cdot 35$. The quantities 5 and 35 are the numbers of markers that the minute hand had to travel to get to the desired positions, assuming the hour hand remained stationary. In other words, these are the hour hand's head starts. Obviously, the hour hand does not remain stationary. Hence, we multiply these quantities by $\frac{12}{11}$ because the minute hand must travel $\frac{12}{11}$ as far. Let us refer to this fraction, $\frac{12}{11}$, as the *correction factor*.

You can justify the correction factor for the interval between overlaps in the following way: Think of the hands of a clock at noon. During the next 12 hours (i.e., until the hands reach the same position at midnight), the hour hand makes one revolution and the minute hand makes 12 revolutions. The minute hand thus coincides with the hour hand 11 times (including midnight, but not noon, starting just after the hands separate at noon). Whereas each hand rotates at a uniform rate, the hands overlap each $\frac{12}{11}$ of an hour, or $65\frac{5}{11}$ min.

The correction factor, $\frac{12}{11}$, is thus the key that makes what seem to be difficult problems quite easy. Multiplying $\frac{12}{11}$ by 5, 10, 15, and so forth, students can quickly calculate the overlap times, $1:05\frac{5}{11}$, $2:10\frac{10}{11}$, $3:16\frac{4}{11}$, and so forth.

### Extension

There are many other interesting, and sometimes rather difficult, problems that are made simple by this correction factor. For example, you may ask your students to find the exact times when the hands of a clock will be perpendicular (or form a straight angle) between, say, 8 and 9 o'clock.

Again, you would have the students determine the number of markers that the minute hand would have to travel from the "12" position until it forms the desired angle with a *stationary* hour hand. Then have them multiply this number by the correction factor to obtain the exact actual time. That is, to find the exact time that the hands of a clock are *first* perpendicular between 8 and 9 o'clock, determine the desired position of the minute hand when the hour hand remains stationary (here, on the 25 minute marker). Then multiply 25 by $\frac{12}{11}$ to get $8:27\frac{3}{11}$, the exact time when the hands are *first* perpendicular after 8 o'clock.

Your students should derive a great sense of achievement and enjoyment as a result of employing this simple procedure to solve what usually appears to be a very difficult clock problem.

# Digit Problems Revisited

If you have already done "Howlers," you have seen that some problems are easy to tackle if you express a two-digit number as $10a + b$. The $a$ stands for the tens' digit and the $b$ represents the units' digit. Similarly, we might express a four-digit number as $1000a + 100b + 10c + d$. Use this to show why a number is divisible by 9 if its digits add up to 9. Use a four-digit number for your example. Keep in mind that $1000a$ is the same as $999a + a$.

_____

_____

_____

Now try this one: Two different two-digit numbers are formed using the same two digits; that is, one number is the reverse of the other. If the numbers are squared and the smaller square is subtracted from the larger, the difference is 7128. If the two numbers are added, the sum is 22 times as large as the difference between the digits. What is the number?

_____

_____

_____

There is a two-digit number that is evenly divisible by 4 and all its positive integral powers end in the same two digits as the original number. In other words, if our number is $x$, then $x^2$, $x^3$, $x^4$, ..., all end in the same two digits.

What is this number?

_____

_____

_____

_____

Finally, seven times a two-digit number is a three-digit number. Adding a 6 to the end creates a four-digit number that is 1833 greater than the three-digit number. What is the two-digit number?

_____

_____

_____

_____

**EXTENSION!** By shifting the initial digit 6 of the positive integer $N$ to the end, we obtain a number equal to $\frac{1}{4}N$. Find the *smallest* possible value of $N$ for which this can be true.

# Teacher's Notes for Digit Problems Revisited

*This activity can be a real eye opener. To many of your average students, these problems will at first appear extremely difficult—well beyond their abilities or ingenuity. Yet after you've walked them through a few of the problems, students realize that they break down to a fairly easy, systematic analysis. After practice with a few more such problems, students can become quite adept with them and this can be a real confidence booster.*

*Students should be able to solve simple systems of simultaneous equations and, although it's not mandatory, it's usually a good idea for them to have worked through "Howlers" as well.*

--------------------------------- NCTM Standards ---------------------------------

| 1 | 2 | 3 | 4 | 5 | 6 | 7 | 8 | 9 | 10 |
|---|---|---|---|---|---|---|---|---|----|
| • | • |   |   |   | • | • | • | • |    |

### Presenting the Activity

As previously mentioned, mastering the problems of this activity can be a real confidence booster. However, although the solutions require no concepts or operations that you haven't already taught, some ingenuity is called on to translate the problems into appropriate simple equations. We caution you not simply to assign this activity as homework; this could only frustrate the majority of your students. Taking them through the problems step by step can be a pleasant and rewarding experience for everyone.

The hint that's given (i.e., that $1000a$ can be expressed as $999a + a$) should make for an easy start to this activity. Students can readily see that if $1000a + 100b + 10c + d$ is written as

$$999a + a + 99b + b + 9c + c + d,$$

we then have

$$999a + 99b + 9c + a + b + c + d.$$

The first three terms of the expression are obviously divisible by 9. And so, if $a+b+c+d$ is divisible by 9, the entire number must be divisible by 9.

For the second problem, begin by asking how you could designate two two-digit numbers that are the reverse of each other: $10t + u$ and $10u + t$. This gives us the first equation,

$$(10t + u)^2 - (10u + t)^2 = 7128,$$
$$100t^2 + 20tu + u^2 - 100u^2 - 20tu - t^2 = 7128,$$
$$99t^2 - 99u^2 = 7128,$$
$$t^2 - u^2 = 72,$$

and this is as far as we can go with this one. For the second equation, we're told that

$$(10t + u) + (10u + t) = 22(t - u),$$
$$11t + 11u7 = 22(t - u),$$
$$t + u = 2(t - u),$$
$$t = 3u.$$

Substituting in the first equation,

$$9u^2 - u^2 = 72,$$
$$u^2 = 9,$$
$$u = 3.$$

Therefore, $t = 9$.

The original numbers are thus 93 and 39. To check: $93^2 - 39^2 = 8649 - 1521 = 7128$, and $93 + 39 = 22(9 - 3) = 132$.

On the next problem, students will probably suggest calling the number $10t + u$, and this is fine. At this point, you're in a position to suggest some shortcuts—eliminating some possibilities. Ask students what they already know about $u$. Because the number is divisible by 4, $u$ must be even. Ask students what even digits have squares that end in the same digit. There are only two: 0 and 6. If 0 is chosen, the only answer that fits the problem's criteria is 00, and so this case is trivial. With some confidence then, we can now assign $u = 6$.

From here on we could simply plug in values of 1–9 for $t$, but we can more easily narrow the possibilities for $t$. Again, because the number is divisible by 4, we can say $10t + 6 = 4m$, where $m$ stands for some integer. Then, $5t + 3 = 2m$, and so $t$ must be odd. Our trial and error now is $16^2 = 256$, reject; $36^2 = 1296$, reject; $56^2 = 3136$, reject; and $76^2 = 5776$, eureka! Have students verify this with successive powers of 76.

By the time students reach the fourth problem they're probably getting into the habit of using $10t + u$. An easier way, however, is to let the two-digit number be $x$. Then the three-digit number is $7x$. Be sure that students realize that the problem asks for a 6 to be tacked onto the end of $7x$ (making it a four-digit number), not just adding 6 to $7x$. Thus, the three digits are shifted one place to the left; i.e., multiplied by 10. Our equation then is $70x + 6 = 7x + 1833$, and $x = 29$.

### Extension

A quick inspection shows that the answer to the problem can't be a two-digit number: when we reverse 61, we find 16 is already greater than one-fourth of 61. Thus, there's no sense in trying $62, 63, \ldots$. When we try a three-digit number, some students may try to use $600 + 10t + u$ to represent $N$. Again, this can't work because we can derive only one equation from the given information. If we substitute $x$ for the last two digits, $N = 600 + x$. In moving the 6 to the end we shift $x$ one place to the left. Therefore, our new number is $10x + 6$ and our equation is

$$\frac{1}{4}(600 + x) = 10x + 6.$$

This yield $39x = 576$, which doesn't divide evenly. Neither do our attempts with four and five digits, but at six digits,

$$\frac{1}{4}(600{,}000 + x) = 10x + 6,$$

$$39x = 599{,}976,$$

$$x = 15{,}384,$$

and thus

$$N = 615{,}384.$$

Check to find that $4 \times 153{,}846 = 615{,}384$.

# CHAPTER 5

# Recreational Mathematics

- Howlers
- Tessellations
- Cyclic Numbers
- The Parabolic Envelope ★

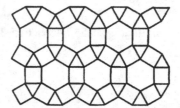

Recreational mathematics has gotten what so many presidential secretaries have complained about: bad press. However, as Martin Gardner put it so well, "A good mathematical puzzle, paradox, or magic trick can stimulate a child's imagination much faster than a practical application (especially if the application is remote from the child's experience), and if the "game" is chosen carefully it can lead almost effortlessly into significant mathematical ideas." For the reasons Gardner cites, these units should not be neglected or considered trivial, "merely recreational." Moreover, they should not be presented as a block, but rather interspersed throughout the year. They're especially valuable to use when interest in general seems to be waning.

"Howlers" is sheer fun for anyone with a mathematical sense of humor. It's one of those odd examples wherein you're able to get the correct answer by performing a very incorrect operation. (Similar howlers can be found in sums of yards, feet, and inches, and the old system of pounds, shillings, and pence.)

"Tessellations" will appeal to those who enjoy geometry and to all jigsaw-puzzle buffs. As with "Algebraic Identities" and several of the activities in *Making Pre-Algebra Come Alive*, the strong link between algebra and geometry is shown. A lot of time should be allowed for both cutting and playing around with the arrangements of the polygons. It's thus best to assign this one as homework or for a free period before discussing it in class. The Extension can be done on a trial-and-error basis by all your students. The algebraic proof that only eight semiregular tessellations can be formed should only be attempted by your top students.

"Cyclic Numbers" is another of those delightful anomalies in our number system that will fascinate those of your students who are already hooked on mathematics, but may produce enormous yawns in the remainder of the unenlightened population. Students who find the strange properties of cyclic numbers appealing will also enjoy "Scamps" in *Making Pre-Algebra Come Alive*.

Most drawings that students are instructed to make don't turn out very well, and it's sometimes discouraging to students when they compare their work with the pretty pictures in the book. "The Parabolic Envelope" offers a refreshing exception—the kids' drawings usually look pretty good.

This activity makes another good link between algebra and geometry, and hidden in this "recreational" topic is an important mathematical skill: Based on what they discover from their measurements of the parabola, students are asked to come up with a valid definition of the parabola. The Alternate Extension makes this activity appropriate for both first- and second-year algebra students.

If your students have not been introduced to the recreational section of *Making Pre-Algebra Come Alive*, these activities will be both challenging and entertaining. The same is true for the binary activities in *Making Pre-Algebra Come Alive*.

# Howlers

Can you reduce the following fractions to their lowest terms?

$$\frac{16}{64} = \underline{\hspace{2cm}}, \qquad \frac{19}{95} = \underline{\hspace{2cm}}.$$

If you used your usual means to reduce fractions, you did a lot of unnecessary work. Look at the Hint at the bottom of the next page.

In *Fallacies in Mathematics*, E. A. Maxwell called these fractions *howlers*. Can you think of other two-digit fractions that can be reduced by cancelling this way? How might we show that there are or are not additional examples?

These fractions can be expressed as $\frac{10x+a}{10a+y}$. The foregoing two cancellations are such that when we cancel the $a$s, the fraction equals $\frac{x}{y}$.

Therefore,

$$\frac{10x + a}{10a + y} = \frac{x}{y},$$
$$y(10x + a) = x(10a + y),$$
$$10xy + ay = 10ax + xy,$$
$$9xy + ay = 10ax,$$
$$y = \frac{10ax}{9x + a}.$$

Keep in mind that $x$ and $a$ must be single-digit integers. Fill in the following chart. Wherever $y$ is an integer, you've got a howler. How many are there? ___

| $x$ \ $a$ | 1 | 2 | 3 | 4 | 5 | 6 | 7 | 8 | 9 |
|---|---|---|---|---|---|---|---|---|---|
| 1 | ▨ | $\frac{20}{11}$ | $\frac{30}{12}$ | $\frac{40}{13}$ | | | | | |
| 2 | $\frac{20}{19}$ | ▨ | $\frac{60}{21}$ | | | | | | |
| 3 | $\frac{30}{28}$ | $\frac{60}{29}$ | ▨ | | | | | | |
| 4 | | | | ▨ | | | | | |
| 5 | | | | | ▨ | | | | |
| 6 | | | | | | ▨ | | | |
| 7 | | | | | | | ▨ | | |
| 8 | | | | | | | | ▨ | |
| 9 | | | | | | | | | ▨ |

**EXTENSION!** Find some fractions with more than two digits in the numerator and denominator that are howlers.

*Hint:* Cancel the digits that are the same in the numerator and denominator.

# Teacher's Notes for Howlers

*In addition to being a lot of fun, working with howlers can provide elementary algebra students an increased understanding of number concepts. Your students should be able to reduce fractions to lowest terms. They should also be familiar with factors and be able to perform all operations on fractions.*

——————————————————— NCTM Standards ———————————————————

| 1 | 2 | 3 | 4 | 5 | 6 | 7 | 8 | 9 | 10 |
|---|---|---|---|---|---|---|---|---|----|
| • | • |   |   |   | • | • | • | • | •  |

### Presenting the Activity

Begin your presentation by asking students to reduce to lowest terms the fractions $\frac{16}{64}$ and $\frac{19}{95}$. After they have reduced the fractions to lowest terms in the usual manner, tell them that they did a lot of unnecessary work. Show them the cancellations (this, of course, assuming that you have blocked out the hint at the bottom of the student page)

$$\frac{1\cancel{6}}{\cancel{6}4} = \frac{1}{4}, \qquad \frac{1\cancel{9}}{\cancel{9}5} = \frac{1}{5}.$$

At this point your students will be somewhat amazed. Their first reaction is to ask if this can be done to any fraction composed of two-digit numbers. Challenge your students to find another fraction (composed of two-digit numbers) where this type of cancellation will work. Students might cite $\frac{55}{55} = \frac{5}{5} = 1$ as an illustration of this type of cancellation. Tell them that although this will hold true for all multiples of 11, it is trivial, and our concern will be only with proper fractions (i.e., whose value is less than 1).

Now discuss why these fractions can be reduced in this manner. The students should realize from the value found for $y$ that it is necessary for $x$, $y$, and $a$ to be integers, because they are digits in the numerator and denominator of a fraction. It is now their task to find the values of $a$ and $x$ for which $y$ will also be integral.

To avoid a lot of algebraic manipulation, students should complete the chart, which will generate values of $y$ from

$$y = \frac{10ax}{9x + a}.$$

Remind them that $x$, $y$, and $a$ must be single-digit integers. The complete table they are to construct follows. Notice that the cases where $x = a$ are excluded because $\frac{x}{a} = 1$.

| $x \backslash a$ | 1 | 2 | 3 | 4 | 5 | 6 | 7 | 8 | 9 |
|---|---|---|---|---|---|---|---|---|---|
| 1 | ▨ | $\frac{20}{11}$ | $\frac{30}{12}$ | $\frac{40}{13}$ | $\frac{50}{14}$ | $\boxed{\frac{60}{15}}$ | $\frac{70}{16}$ | $\frac{80}{17}$ | $\boxed{\frac{90}{18}}$ |
| 2 | $\frac{20}{19}$ | ▨ | $\frac{60}{21}$ | $\frac{80}{22}$ | $\frac{100}{23}$ | $\boxed{\frac{120}{24}}$ | $\frac{140}{25}$ | $\frac{160}{26}$ | $\frac{180}{27}$ |
| 3 | $\frac{30}{28}$ | $\frac{60}{29}$ | ▨ | $\frac{120}{31}$ | $\frac{150}{32}$ | $\frac{180}{33}$ | $\frac{210}{34}$ | $\frac{240}{35}$ | $\frac{270}{36}$ |
| 4 | $\frac{40}{37}$ | $\frac{80}{38}$ | $\frac{120}{39}$ | ▨ | $\frac{200}{41}$ | $\frac{240}{42}$ | $\frac{280}{43}$ | $\frac{320}{44}$ | $\boxed{\frac{360}{45}}$ |
| 5 | $\frac{50}{46}$ | $\frac{100}{47}$ | $\frac{150}{48}$ | $\frac{200}{49}$ | ▨ | $\frac{300}{51}$ | $\frac{350}{52}$ | $\frac{400}{53}$ | $\frac{450}{54}$ |
| 6 | $\frac{60}{55}$ | $\frac{120}{56}$ | $\frac{180}{57}$ | $\frac{240}{58}$ | $\frac{300}{59}$ | ▨ | $\frac{420}{61}$ | $\frac{480}{62}$ | $\frac{540}{63}$ |
| 7 | $\frac{70}{64}$ | $\frac{140}{65}$ | $\frac{210}{66}$ | $\frac{280}{67}$ | $\frac{350}{68}$ | $\frac{420}{69}$ | ▨ | $\frac{560}{71}$ | $\frac{630}{72}$ |
| 8 | $\frac{80}{73}$ | $\frac{160}{74}$ | $\frac{240}{75}$ | $\frac{320}{76}$ | $\frac{400}{77}$ | $\frac{480}{78}$ | $\frac{560}{79}$ | ▨ | $\frac{720}{81}$ |
| 9 | $\frac{90}{82}$ | $\frac{180}{83}$ | $\frac{270}{84}$ | $\frac{360}{85}$ | $\frac{450}{86}$ | $\frac{540}{87}$ | $\frac{630}{88}$ | $\frac{720}{89}$ | ▨ |

When the chart on the student page is completed it will show four integral values of $y$:

When $x = 1$ and $a = 6$, then $y = 4$.
When $x = 2$ and $a = 6$, then $y = 5$.
When $x = 1$ and $a = 9$, then $y = 5$.
When $x = 4$ and $a = 9$, then $y = 8$.

These values yield the fractions $\frac{16}{64}$, $\frac{26}{65}$, $\frac{19}{95}$, and $\frac{49}{98}$, respectively. This exhausts all possibilities and should convince students that there are only four such fractions composed of two-digit numbers.

*Extension*

Students may now wonder if there are fractions composed of numerators and denominators of more than two digits where this strange type of cancellation can be employed. Have students try this type of cancellation with $\frac{499}{998}$. They should find that, in fact, $\frac{499}{998} = \frac{4}{8} = \frac{1}{2}$. Soon they will realize that

$$\frac{49}{98} = \frac{499}{998} = \frac{4999}{9998} = \frac{49999}{99998} = \cdots,$$

$$\frac{16}{64} = \frac{166}{664} = \frac{1666}{6664} = \frac{16666}{66664} = \frac{166666}{666664} = \cdots,$$

$$\frac{19}{95} = \frac{199}{995} = \frac{1999}{9995} = \frac{19999}{99995} = \frac{199999}{999995} = \cdots,$$

$$\frac{26}{65} = \frac{266}{665} = \frac{2666}{6665} = \frac{26666}{66665} = \frac{266666}{666665} = \cdots.$$

Students who have a further desire to seek additional fractions that permit this strange cancellation should be shown the fractions

$$\frac{332}{830} = \frac{32}{80} = \frac{2}{5}, \qquad \frac{385}{880} = \frac{35}{80} = \frac{7}{16}, \qquad \frac{138}{345} = \frac{18}{45} = \frac{2}{5},$$

$$\frac{275}{770} = \frac{25}{70} = \frac{5}{14}, \qquad \frac{163}{326} = \frac{1}{2}, \qquad \frac{203}{609} = \frac{1}{3}.$$

They should verify the legitimacy of this cancellation and then set out to discover more such fractions.

# Tessellations

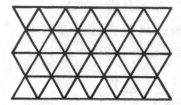

Equilateral triangles can be fitted together to cover a flat surface as shown in the preceding diagram. What other regular polygons can be fitted together this way?

_____

When polygons are fitted together to cover a plane with no spaces between them and no overlapping, the pattern is a *tessellation*. In the preceding tessellation, the polygons are regular polygons: they are all the same size and shape, and no vertex of one polygon meets a side of another. This pattern is a *regular tessellation*. The arrangement of the polygons at each vertex is the same. For the equilateral triangles, six three-sided polygons meet at each vertex. We can write this as 3-3-3-3-3-3.

Cut out the polygons on the next page. Which regular polygons can form a tessellation like the triangles? _____

Write these tessellations using the notation at the end of the second paragraph:

_____

Tessellations can also be formed by fitting together two or more different kinds of regular polygons. The following figure shows a tessellation of hexagons and triangles.

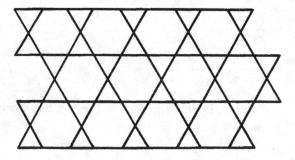

Write the arrangement of the polygons at each vertex using the notation at the end of the second paragraph: _____

To find other tessellations, first experiment to see how many ways polygons can be fitted together at a vertex. Using the polygons you cut out, write the ways three of these polygons can be arranged at a vertex:

_____  _____  _____  _____

Write the ways four of the polygons can be arranged at a vertex:

_____  _____  _____

_____  _____  _____

Write the ways five of the polygons can be arranged:

_____  _____  _____

Write the ways six of the polygons can be arranged: _____

Can you fit more than six of the polygons at a vertex? _____

Can you fit two polygons at a vertex? _____

Extend each of the preceding vertex arrangements to form a tessellation. Sketch the tessellations on a separate sheet of paper.

**EXTENSION!** The tessellation that uses hexagons and triangles is a *semiregular tessellation*. In a semiregular tessellation, the arrangement of the polygons must be the same at every vertex. Which of the tessellations you found previously are semiregular?

# Teacher's Notes for Tessellations

*This activity is an excellent change-of-pace activity that can be considered mainly recreational or can be expanded as subsequently described to include the mathematical justification for the number of regular and semiregular tessellations. It also provides a good link between algebraic and geometric concepts. Reproducible polygons to be used in "Tessellations" are given on the next page.*

|          |          |          | NCTM Standards |   |   |   |   |   |    |
|----------|----------|----------|----------|----------|----------|----------|----------|----------|----------|
| 1        | 2        | 3        | 4        | 5        | 6        | 7        | 8        | 9        | 10       |
| •        | •        | •        |          |          |          |          |          |          |          |

### Presenting the Activity

There are many different kinds of tessellations. Some, such as linoleum patterns, use geometric shapes as discussed in this activity. Others, such as many of the works of Maurits Escher, use pictures to tessellate a plane. If possible, have pictures of various tessellations to show the class.

Several students will realize that squares and hexagons can also form regular tessellations. By experimenting with the cut-out polygons, all students will find these two regular tessellations. They are written as 4-4-4-4 and 6-6-6.

Some students may wonder whether regular polygons that are not included on the extra page can form regular tessellations. Explain that for the polygons to fit together exactly, the sum of the angles at each vertex must be exactly 360°. Also, there must be at least three polygons at a vertex. Thus, the angle of the polygon can be no more than $360 \div 3$ or 120°. A pentagon cannot be used to form a regular tessellation because three pentagons form an angle sum of 324° and four pentagons would be 432°. There are only three regular tessellations.

| Number of Sides | Measure of an Angle |
|-----------------|---------------------|
| 3               | 60°                 |
| 4               | 90°                 |
| 5               | 108°                |
| 6               | 120°                |
| 7               | $128\frac{4}{7}°$   |
| 8               | 135°                |
| ⋮               | ⋮                   |
| $n$             | $\frac{(n-2)180}{n}$ |

Next, students experiment with tessellations using two or more kinds of polygons. First, they consider vertex arrangements. Emphasize that the vertex notation must be written either clockwise or counterclockwise around the vertex. Thus, 3-3-6-6 is a different arrangement than 3-6-3-6. There are four ways that three polygons can be arranged at a vertex: 6-6-6, 4-6-12, 4-8-8, and 3-12-12. There are seven ways for four polygons: 4-4-4-4, 3-3-4-12, 3-4-3-12, 3-3-6-6, 3-6-3-6, 3-4-4-6, and 3-4-6-4. There are three ways for five polygons: 3-3-3-3-6, 3-3-3-4-4, and 3-3-4-3-4. There is one way for six polygons: 3-3-3-3-3-3. There can be no more than six polygons at a vertex because there is no polygon with fewer sides than a triangle and only six triangles will fit at a vertex.

Tessellations formed from the foregoing vertex arrangements are given on the back of the next page. (Students may discover additional tessellations from these vertex arrangements.) The semiregular tessellations are discussed in the Extension.

### *Extension*

There are eight semiregular tessellations. They are shown on page 106 with their vertex arrangements. The three regular tessellations are also shown, leaving four vertex arrangements that do not form semiregular tessellations. When these vertex arrangements are extended to form tessellations, the arrangement of the polygons is *not* the same at every vertex.

There are other regular polygons that can be arranged at a vertex. These can be found by considering a vertex angle of a regular $n$-gon. This vertex angle measures

$$\frac{(n-2)180}{n} \quad \text{or} \quad 180\left(1 - \frac{2}{n}\right).$$

If there are three regular polygons of $n_1$, $n_2$, and $n_3$ sides, then

$$180\left(1 - \frac{2}{n_1}\right) + 180\left(1 - \frac{2}{n_2}\right) + 180\left(1 - \frac{2}{n_3}\right) = 360,$$

$$1 - \frac{2}{n_1} + 1 - \frac{2}{n_2} + 1 - \frac{2}{n_3} = 2,$$

$$\frac{2}{n_1} + \frac{2}{n_2} + \frac{2}{n_3} = 1,$$

$$\frac{1}{n_1} + \frac{1}{n_2} + \frac{1}{n_3} = \frac{1}{2}.$$

Similarly, for four, five, and six polygons at a vertex,

$$\frac{1}{n_1} + \frac{1}{n_2} + \frac{1}{n_3} + \frac{1}{n_4} = 1,$$

$$\frac{1}{n_1} + \frac{1}{n_2} + \frac{1}{n_3} + \frac{1}{n_4} + \frac{1}{n_5} = \frac{3}{2},$$

$$\frac{1}{n_1} + \frac{1}{n_2} + \frac{1}{n_3} + \frac{1}{n_4} + \frac{1}{n_5} + \frac{1}{n_6} = 2.$$

The following table shows the 17 possible integral solutions.

| Number | $n_1$ | $n_2$ | $n_3$ | $n_4$ | $n_5$ | $n_6$ |
|--------|-------|-------|-------|-------|-------|-------|
| 1 | 3 | 7 | 42 | | | |
| 2 | 3 | 8 | 24 | | | |
| 3 | 3 | 9 | 18 | | | |
| 4 | 3 | 10 | 15 | | | |
| 5 | 3 | 12 | 12 | | | |
| 6 | 4 | 5 | 20 | | | |
| 7 | 4 | 6 | 12 | | | |
| 8 | 4 | 8 | 8 | | | |
| 9 | 5 | 5 | 10 | | | |
| 10 | 6 | 6 | 6 | | | |
| 11 | 3 | 3 | 4 | 12 | | |
| 12 | 3 | 3 | 6 | 6 | | |
| 13 | 3 | 4 | 4 | 6 | | |
| 14 | 4 | 4 | 4 | 4 | | |
| 15 | 3 | 3 | 3 | 4 | 4 | |
| 16 | 3 | 3 | 3 | 3 | 6 | |
| 17 | 3 | 3 | 3 | 3 | 3 | 3 |

These solutions (and rearrangements of solutions 11–16) all have been considered, with the exception of numbers 1, 2, 3, 4, 6, and 9. These six solutions each can be formed at a single vertex, but they cannot be extended to cover the whole plane. Thus, there are exactly eight semiregular tessellations.

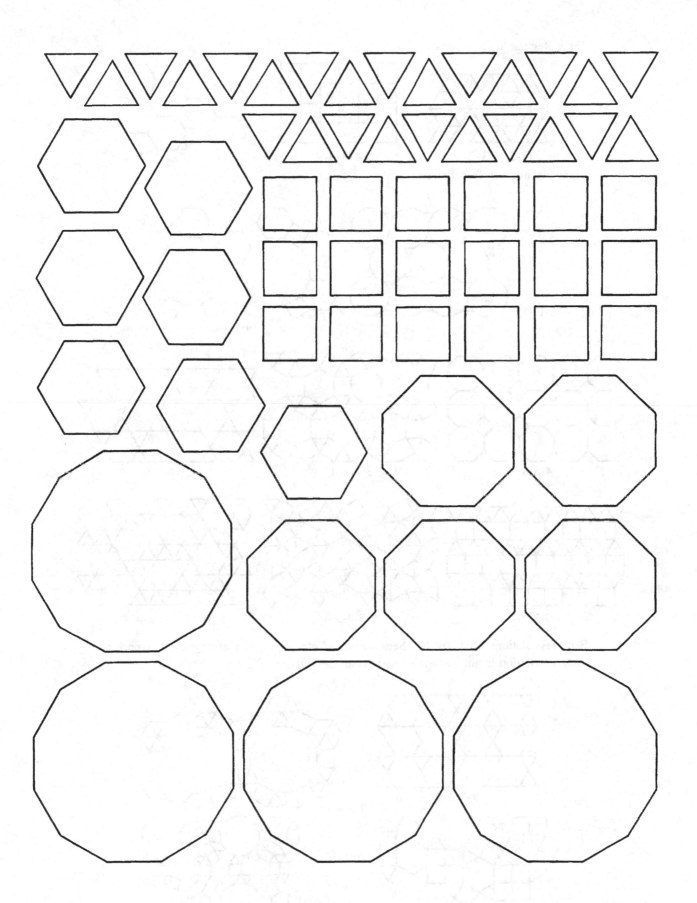

Three Regular Tessellations:

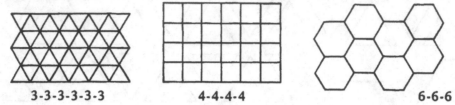

**3-3-3-3-3-3**          **4-4-4-4**          **6-6-6**

Eight Semiregular Tessellations:

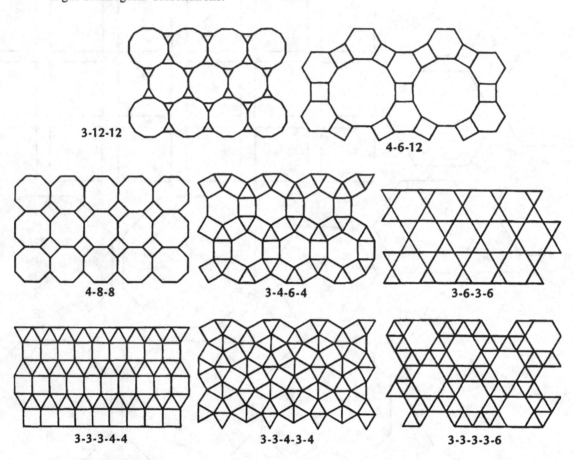

**3-12-12**

**4-6-12**

**4-8-8**          **3-4-6-4**          **3-6-3-6**

**3-3-3-4-4**          **3-3-4-3-4**          **3-3-3-3-6**

Four Tessellations That Are Not Semiregular: (*Note*: The vertex arrangement given for each tessellation is not the arrangement at all the vertices.)

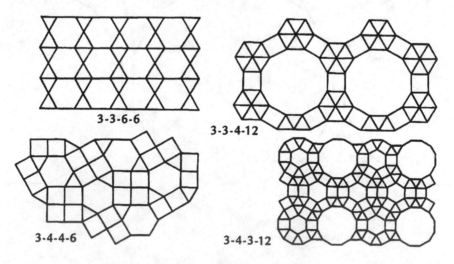

**3-3-6-6**

**3-3-4-12**

**3-4-4-6**

**3-4-3-12**

# Cyclic Numbers

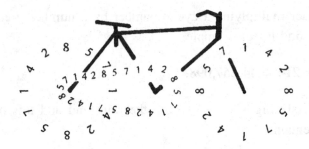

Find the products

$1 \times 142,857 =$ _____,

$2 \times 142,857 =$ _____,

$3 \times 142,857 =$ _____,

$4 \times 142,857 =$ _____,

$5 \times 142,857 =$ _____,

$6 \times 142,857 =$ _____.

What do you notice about your answers? _____

_____

Numbers such as 142,857 are called *cyclic numbers*. Prime numbers and the decimal equivalents of their reciprocals can be used to find cyclic numbers. Let's consider the decimal equivalents of the reciprocals of 2, 3, 5, and 7 (the first four prime numbers). Find the decimal equivalents of

$$\frac{1}{2} = \text{\_\_\_\_} , \qquad \frac{1}{3} = \text{\_\_\_\_} , \qquad \frac{1}{5} = \text{\_\_\_\_} , \qquad \frac{1}{7} = \text{\_\_\_\_} .$$

Only $\frac{1}{3}$ and $\frac{1}{7}$ have a *repetend* (the part of the decimal that repeats). The repetend for $\frac{1}{7}$ is 142,857, the cyclic number at the beginning. This is the smallest cyclic number. The repetend for $\frac{1}{7}$ has six digits, one less than the prime number 7. Thus, for a prime number $\rho$, if $\frac{1}{\rho}$ has $\rho - 1$ digits in its repetend, then the repetend is a cyclic number.

The next larger cyclic number has 16 digits. What prime number is used to find it? _____ What is this 16-digits cyclic number? _____

Cyclic numbers have many interesting properties. Suppose 142,857 is split into 142 and 857. What do you get when you add $142 + 857$? _____ Split the 16-digit cyclic number into two 8-digit numbers and add them.

What do you get? _____ Does this mean cyclic numbers are divisible by 9? _____ Why or why not? _____

Multiply each cyclic number by the prime used to find it. What do you get? __
_____

Finally, consider multiplying a cyclic number by a number greater than $\rho$, the prime used to find it; for example,

$$142,857 \times 214 = 44,857,098.$$

Now mark off six digits $(\rho - 1)$ from the right and add this number and the number that remains:

$$857,098 + 44 = \underline{\hspace{2cm}}.$$

Try $142,857 \times 231$. What do you get? _____
Is 231 a multiple of $\rho$? _____

Try $142,857 \times 703$ and $142,857 \times 84$: _____

**EXTENSION!** Find the next largest cyclic number.

# Teacher's Notes for Cyclic Numbers

*Cyclic numbers are one of the uncanniest occurrences in all of arithmetic. Students who are intrigued by their strange workings will enjoy Martin Gardner's much fuller treatment in Mathematical Circus (Random House, New York, 1981).*

*The concepts involved in this activity are simple. However, a lot of computation is required, so urge students to be careful in their work.*

――――――――――――――――― NCTM Standards ―――――――――――――――――

| 1 | 2 | 3 | 4 | 5 | 6 | 7 | 8 | 9 | 10 |
|---|---|---|---|---|---|---|---|---|----|
| • |   |   |   |   | • |   |   |   |    |

### Presenting the Activity

Students should be surprised when they find the answers to the multiplication problems:

$$1 \times 142{,}857 = 142{,}857,$$
$$2 \times 142{,}857 = 285{,}714,$$
$$3 \times 142{,}857 = 428{,}571,$$
$$4 \times 142{,}857 = 571{,}428,$$
$$5 \times 142{,}857 = 714{,}285,$$
$$6 \times 142{,}857 = 857{,}142.$$

Each answer contains the same digits in the same *cyclic* order. That is, if the digits 142857 are thought of as being connected in a circle, the circle can be broken in six places to produce the preceding six numbers. Hence the name cyclic numbers.

The decimal equivalents of the reciprocals of the first four prime numbers are

$$\frac{1}{2} = 0.5, \qquad \frac{1}{3} = 0.\overline{3}, \qquad \frac{1}{5} = 0.2, \qquad \frac{1}{7} = 0.\overline{142857}.$$

Thus, cyclic numbers can be found from the decimal equivalents of the reciprocals of prime numbers. Emphasize that the repetend must have $\rho - 1$ digits, where $\rho$ is the original prime number.

Whereas the next larger cyclic number has 16 digits, the prime used to find it must be 17. Thus, to find this cyclic number, students must find the decimal equivalent of $\frac{1}{17}$. Although this division is somewhat tedious, it is the least complicated way to find cyclic numbers. The 16-digit number is

0,588,235,294,117,647.

Note that this cyclic number begins with a zero. All cyclics found from primes greater than 7 begin with one or more zeros. Obviously, this is true because all decimal equivalents of fractions of the form $\frac{1}{a}$, where $a > 10$, will have zeros in at least the tenth's place.

Students can multiply the 16-digit cyclic by a few of the numbers between 1 and 17 to check that it is indeed cyclic. These multiplications will also emphasize the need for the zero.

When a cyclic number is split into two numbers of equal length, the sum of these two numbers is a number consisting only of 9s:

$$142 + 857 = 999;$$
$$05,882,352 + 94,117,647 = 99,999,999.$$

Thus, every cyclic number must be divisible by 9 because the sum of its digits is divisible by 9. Note that the same result occurs with any of the permutations of the cyclic number (428,571 and 285,714 are examples of permutations of 142,857).

When a cyclic number is multiplied by the prime used to find it, the result is also a number that consists only of 9s. This provides another method for finding cyclic numbers. Simply divide a prime number, $\rho$, into a row of 9s until there is no remainder. If there are $\rho - 1$ digits in the quotient, then the quotient is a cyclic number.

When one of the permutations of the cyclic number is multiplied by the prime number used to find it, the result is also interesting. In this case, the first and last digits add to 9 and the other digits are all 9s.

Thus far, students have considered what products occur when a cyclic number is multiplied by numbers less than $\rho$ (the generating prime number) and by $\rho$. Next, they see what happens when the cyclic is multiplied by numbers greater than $\rho$.

The product of $142,857 \times 214$ is $44,857,098$. This product is partitioned in $(\rho - 1)$ groups beginning at the right and then the groups are added:

```
  857,098
+      44
  857,142.
```

The result is one of the permutations of the cyclic number.

For $142,857 \times 231$, students will get $32,999,967$. In this case the sum after partitioning is a row of 9s: $999,999$. The multiplier 231 is a multiple of 7, the generating prime. Thus, multiplying a cyclic number by a number greater than $\rho$, and partitioning and adding, results in two possibilities: If the multiplier is *not* a multiple of $\rho$, the product reduces to a permutation of the cyclic number. If the multiplier *is* a multiple of $\rho$, the product reduces to a number consisting of $(\rho - 1)$ nines.

The other products on the student page are

$$142,857 \times 703 = 100,428,471, \quad \begin{array}{r} 428,471 \\ +\ \ \ 100 \\ \hline 428,571, \end{array}$$

$$142,857 \times 84 = 11,999,988, \quad \begin{array}{r} 999,988 \\ +\ \ \ \ 11 \\ \hline 999,999. \end{array}$$

To reduce computation time, remind students to use the products they found at the beginning of the activity.

*Extension*

The next largest cyclic number is generated from the next largest prime number, 19. It is

052,631,578,947,368,421.

There are nine prime numbers less than 100 that generate cyclic numbers. They are 7, 17, 19, 23, 29, 47, 59, 61, and 97.

# The Parabolic Envelope

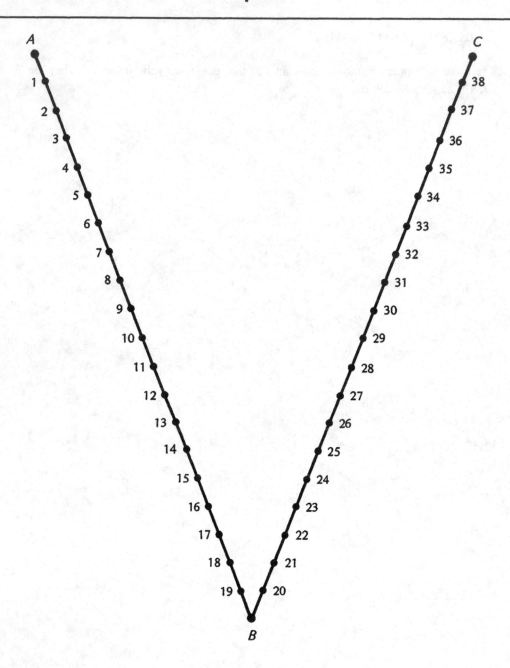

Do you think it's possible to draw a parabola without using a curved line? To try it, let's begin by drawing an angle with line segments of the same length. Then divide each line segment into an *even* number of segments, all of them the same length. Then number the division points as shown in the accompanying figure.

Draw a line segment from point 1 to point 20. (We can call this line segment "1–20.") Now draw 2–21, 3–22, ..., 19–38.

The line segments you have drawn *envelope* a parabola. Measure to find the point in the middle of 10–29. Label this point V. What is point V called? ____

_____

Draw the line from B to V, extending it beyond V. What is this line called? _

_____

What would happen if you folded your paper along this line? _____

_____

Let's find out some properties of parabolas. You will need a protractor to draw a 90° angle. One side of the angle is the line from A to point 10. The other side of the angle goes from point 10 to the line BV. Label the point where the side of the angle intersects line BV, point F. Point F is the *focus* of the parabola.

Next, find the length of the line segment from F to V. Mark a point G below V on line BV so that the distance from G to V is the same as the distance from F to V. Draw a horizontal line through G. This line and 10–29 should be parallel.

The line through G is the *directrix* of the parabola. Choose any point P on the parabola. Measure the distance from P to F. Then measure the *vertical* distance from P to the directrix. What do you notice? _____

Choose another point on the parabola and measure the same way you did before. Is your result the same? _____

Try a few more points.

***EXTENSION!*** Write a definition of a parabola using what you have learned about the focus and the directrix.

# Teacher's Notes for The Parabolic Envelope

*Most algebra students are unaware of the close relationship of algebra topics to geometry. This activity presents students with a method of drawing a parabola without point-by-point plotting from an equation. In addition, the activity extends the ideas usually presented in first-year algebra to include a discussion of the focus and directrix of a parabola. In the Extension, students must write a definition of a parabola using what they have learned about the focus and the directrix.*

*Students will need protractors and rulers to complete the activity and should have completed the parabola discussion in their text.*

### NCTM Standards

| 1 | 2 | 3 | 4 | 5 | 6 | 7 | 8 | 9 | 10 |
|---|---|---|---|---|---|---|---|---|----|
| • | • | • |   |   |   |   |   |   |    |

### Presenting the Activity

Emphasize to students that they must draw their segments carefully if they are to get satisfying results. After having tried to draw parabolas accurately using point-plotting methods, students will be pleased and amazed by the appearance of the parabolic envelope.

Point $V$ is the vertex of the parabola. (Students may call the vertex the "lowest point" or the "minimum point," depending on the definition given in their text.) The segment from $B$ to $V$, extended beyond $V$, is the parabola's axis of symmetry. Students should realize that by folding their paper along this ray, the two halves of the parabola will coincide.

The remainder of the activity will be new to most students and they may need some guidance to find the focus and the directrix. If they have had very little basic geometry in pre-algebra courses, they may need explanations for some of the terms. It is best, however, to first allow students to attempt to find the focus and directrix on their own. Their completed drawings should look like this:

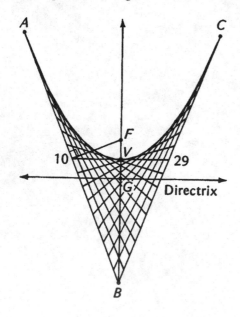

If students have made their drawings carefully, they will find the distance from a point on the parabola to the focus equals the minimum distance from the point to the directrix. (Be sure students understand they must measure the *perpendicular* to the directrix from the point.) After trying a few more points, they should be convinced that these two distances are always the same.

Discuss how the original angle $ABC$ influences the shape of the parabola. If time permits, have students construct parabolic envelopes for different size angles. Then they can find the focus and directrix for these parabolas and verify that the focus and directrix are equidistant from every point of the parabola.

### Extension

Student definitions will vary, but should include the idea of a parabola as the set of all points equidistant from a point (the focus) and a line (the directrix) not containing the point. This, of course, is the definition of a parabola that students learn in second-year algebra.

### Alternate Extension

If students have had a more extensive geometry background, you may want to present the *evolute* to the parabola:

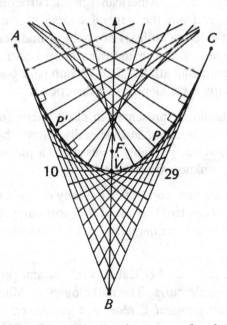

To draw the evolute, students must first locate the points of tangency of the segments used to draw the envelope. To locate these points: (1) Find the point where a segment, say 4–23, intersects the axis of symmetry and label it $T$. (2) Locate $T'$ on the axis above $V$ such that $TV = VT'$. (3) Draw a line through $T'$ parallel to 10–29 so that it intersects the envelope at $P$ and $P'$. $P$ and $P'$ are the points on the parabola where 4–23 and 16–35 are tangent.

Students then draw perpendiculars at each of the points of tangency. These perpendiculars are called *normals*. The envelope of the normals defines the evolute to the parabola.

115

# CHAPTER 6

# Logic

- Venn Diagrams I
- Venn Diagrams II
- I Always Lie
- My Cat Rover
- Algebraic Fallacies

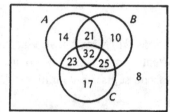

Perhaps the most regrettable change in the American school curriculum this century is the disappearance of logic from the general course of study. This has decreased our students' problem-solving abilities not only in traditional mathematics and physical science applications, but in the even more pervasive areas of politics and consumer and environmental affairs. This section isn't going to cure this situation of course, but perhaps it's a step in that direction.

The two Venn diagram activities should be presented in close succession. "Venn Diagrams I" reviews set notation for students and generally allows them to get their feet wet. The Extension, however, probably gives them a pleasant first taste of using Venn diagrams to find unknown quantities.

"Venn Diagrams II" does not introduce new concepts. It simply extends the ideas of "Venn Diagrams I" so that students can tackle some of the most enjoyable problems they'll encounter in secondary-level mathematics. The Extension is particularly tantalizing.

The next two activities, "I Always Lie" and "My Cat Rover," present problems that are favorites of logicians and puzzle buffs. The anthologies of Martin Gardner (*Mathematical Magic Show, Mathematical Circus,* and several others) and the past several years of *Games* magazine provide literally hundreds of similar problems that are cleverly and entertainingly presented. Although there are answer keys, their problems aren't broken down to step-by-step instructional materials. Thus, they're out of reach for all but top-echelon high-school students, and only mathematically oriented students would search them out. Practice with a few problems such as these activities gives students an opportunity to take advantage of these excellent publications.

The truth-teller/liar problems given in "I Always Lie" are always entertaining. More importantly, they are good exercises in logical problem solving. In the classes I've tested, these problems invariably lead to lively class discussion and can be used as a fine change-of-pace activity anywhere in the course.

"My Cat Rover" usually overwhelms people. Their initial reaction is, "I can't sort out all of this." However, the puzzle is so interesting to most students that they plug away at it. By taking things one step at a time, they're usually able to complete the matrix.

"Algebraic Fallacies" may be presented any time after students have been exposed to squares and square roots. If you wish to use this activity before imaginary numbers have been covered, simply omit the Extension.

# Venn Diagrams I

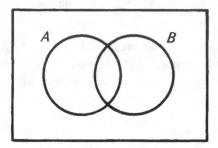

You have probably seen Venn diagrams. They were invented about 100 years ago by an English clergyman and logician, John Venn. He originally devised these diagrams to help illustrate classical Greek logic propositions, but as we shall see, Venn diagrams are very useful in helping to see the relationships among many sets of objects and numbers.

These simple diagrams can be used to get across a lot of information in a very small amount of space. Let's call one class of objects *A* and the other *B*. If we want to refer to something that can be part of *either A or B*, we refer to the *union* of *A* and *B*. This is written $A \cup B$. Something that belongs to *both A and B* is the *intersection* of *A* and *B*, written $A \cap B$. Everything not included in a set is called the *complement* of the set. Thus, everything that isn't in set *A* is called $A^c$. Use this notation to identify the shaded areas in the following figures:

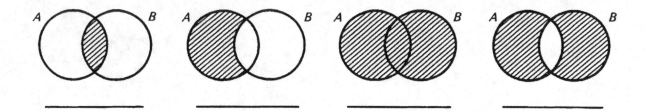

Notice the Venn diagram

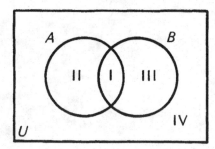

The rectangle represents the universal set, *U*. It's assumed that *A* and *B* are subsets of *U*. If *U* equals all high school students, *A* equals students who take algebra, and *B* equals students who take French, what do regions I–IV represent (what kind of people)?

I = _____ ,

II = _____ ,

III = _____ ,

IV = _____ .

If *U* equals the set of all people, *A* equals all males, and *B* equals all adults, what do regions I–IV represent?

I = _____ ,

II = _____ ,

III = _____ ,

IV = _____ .

Now suppose *U* equals all living things, *A* equals all people, and *B* equals all things with feathers. What does region I represent? _____

***EXTENSION!*** A group of 100 student athletes take part in a test to see if certain vitamins improve their performance. Vitamin A is given to 65 students. Vitamin B is given to 75 students. In talking among themselves they learn that 50 students are getting both vitamins A and B. How many are getting no vitamins?

# Teacher's Notes for Venn Diagrams I

*Most students are given at least some introduction to Venn diagrams before eighth grade. Unfortunately, these introductions usually stop short of showing that Venn diagrams can easily be used to discover quantitative relationships among sets. This activity and the following one give a series of problems that use quantitative methods and that most students find entertaining.*

| | | | | NCTM Standards | | | | | |
|---|---|---|---|---|---|---|---|---|---|
| 1 | 2 | 3 | 4 | 5 | 6 | 7 | 8 | 9 | 10 |
| • | • | | | | • | • | • | • | |

### Presenting the Activity

The activity begins with a review of standard set notation and terminology. Students are asked to recall that $A \cup B$ is read "the union of sets $A$ and $B$." This set contains all elements that belong to *either* $A$ or $B$. $A \cap B$ is read "the intersection of sets $A$ and $B$." This set contains all elements that are members of *both* sets $A$ and $B$. Finally, students are given that $A^c$ equals the complement of set $A$, that is, all elements that are not a part of set $A$. Students can then demonstrate their grasp of this notation by identifying the diagrams:

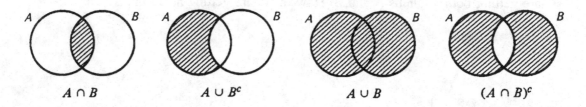

$$A \cap B \qquad A \cup B^c \qquad A \cup B \qquad (A \cap B)^c$$

The activity then shows how to apply the notation to concrete situations. In the next Venn diagram, region I is obviously the students who study *both* algebra and French. Region II indicates the students who are taking algebra but not French. Region III represents the students who are taking French but not algebra. Region IV is, of course, students who are taking neither.

Most students find the next Venn diagram a more interesting puzzle. Region I is men—both male and adult; region II is boys—male, but not adult; region III is women—adult, but not male; and region IV is girls—neither male nor adult. The last question allows the introduction and discussion of the empty or null set.

### Extension

Students should have no difficulty solving the Extension if they use a Venn diagram exactly like the one given at the top of the first student page. Region I is given as 50 students. Thus, region II must be $65 - 50 = 15$ students. Region III is $75 - 50 = 25$ students. Added together, regions I, II, and III account for 90 students, so region IV must account for 10 students.

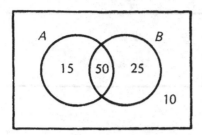

### Alternate Extension

You may wish to give your students an additional problem as practice before proceeding to "Venn Diagrams II":

A class of 30 students have 70 pets: 45 dogs and 25 cats. Eight of the students say they have one or more dogs, but no cats. Eight other students say they have no pets at all. One boy claims to have 10 cats and 1 browbeaten dog, and a girl says she has just the opposite: 10 dogs and 1 very nervous cat. Six other students say they also have both dogs and cats, but not in those extreme proportions. Fill in a Venn diagram to show how many students have dogs, cats, both, and neither.

Such a problem is both more realistic and more challenging because it stirs up some irrelevant dust: The numbers of pets are beside the point. All that is asked for is the number of students owning which kind of pet. If students keep their eyes on this question, they have no trouble producing the following Venn diagram:

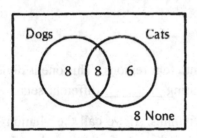

# Venn Diagrams II

You have already seen that Venn diagrams can be used to classify numbers or objects. You've also seen that they can assist in finding how many elements are in each set. This activity will show you how to use Venn diagrams to help with more complicated problems. Look at the following Venn diagram:

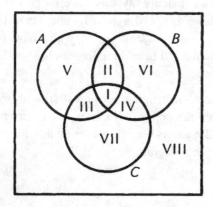

The last Venn diagram you saw was divided into four regions. This one is divided into _____ regions because we are comparing _____ different sets.

Let's return to the problem of athletes and vitamins. If we call the vitamins $A$, $B$, and $C$, then region I, $(A \cap B \cap C)$, describes the set of athletes who take all three vitamins. Region III, $(A \cap B^c \cap C)$, is the set of all athletes who take $A$ and $C$, but not $B$.

Using the following Venn diagrams, shade the area and give the set notation for (1) athletes who take *at least* one kind of vitamin and (2) athletes who take *only* one kind of vitamin.

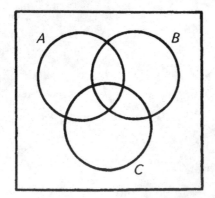 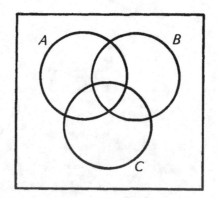

Suppose that out of a total group of 150 athletes, it is found that

90 take vitamin $A$,     53 take both vitamin $A$ and $B$,

88 take vitamin $B$,     55 take both vitamin $A$ and $C$,

97 take vitamin $C$,     57 take both vitamin $B$ and $C$,

32 take all three vitamins.

How many take none of the three vitamins?

Look at the first Venn diagram on the previous page. Because 32 athletes take all three vitamins, place this number in region I ($A \cap B \cap C$). Because 53 athletes take both $A$ and $B$, and of these 32 also take $C$, there must be $53 - 32 = 21$ who take $A$ and $B$, but not $C$. Similarly, _____ athletes take $A$ and $C$, but not $B$ (region III). _____ athletes take $B$ and $C$, but not $A$ (region IV).

We have now accounted for $32 + 21 + 23 = 76$ of the 90 athletes who take vitamin $A$. This leaves $90 - 76 = 14$ who take $A$, but not $B$ or $C$.

So, _____ athletes take $B$ alone and _____ athletes take $C$ alone.

Now we can calculate that _____ athletes take none of the vitamins.

*EXTENSION!* Last month Congressman Flipflop hired an opinion research firm to survey voter attitudes in his district. The survey results included the following figures: 80% of the voters interviewed favor gun control, 53% favor nuclear power, and 69% favor tighter pollution limits. Both gun control *and* nuclear power were favored by 21% of those surveyed. Likewise, 46% favored both gun control and pollution limits, 34% favored both nuclear power and pollution limits, and 9% favored all three. After studying the report overnight, the congressman accused the research firm of not knowing what they were doing. He said their survey results were impossible. How did he reach this conclusion?

# Teacher's Notes for Venn Diagrams II

*This activity picks up where "Venn Diagrams I" leaves off. It has the motivational advantage of presenting problems that at first glance appear very difficult, but that break down to easily handled exercises. Students should have completed "Venn Diagrams I" before studying this activity.*

─────────── NCTM Standards ───────────

| 1 | 2 | 3 | 4 | 5 | 6 | 7 | 8 | 9 | 10 |
|---|---|---|---|---|---|---|---|---|----|
|   | • | • |   |   | • | • | • | • |    |

### Presenting the Activity

The Venn diagrams in this activity specify three sets and thus contain eight regions. Review the set notation from "Venn Diagrams I" before tackling the problems of this activity. The answers for athletes who take at least one kind of vitamin and only one kind of vitamin are

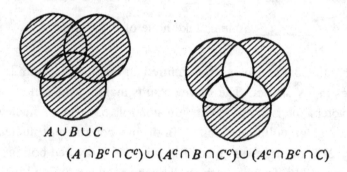

$$A \cup B \cup C$$

$$(A \cap B^c \cap C^c) \cup (A^c \cap B \cap C^c) \cup (A^c \cap B^c \cap C)$$

The set notation in the second case is about as complex as it can be, so caution students not to be discouraged if they have difficulty here. You can ease the problem somewhat by having students first define "students who take only vitamin A." This, of course is simply $A \cap B^c \cap C^c$.

The completed Venn diagram for the vitamin problem is

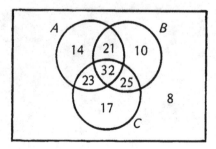

All of these figures are derived by simple subtraction.

Students work backward from the given information. Because 32 athletes take all three vitamins, 32 is placed in region I as discussed on the student pages. Next, the cases are considered for athletes who take two vitamins but not the third. Because 53 athletes take both $A$ and $B$, and of these, 32 also take $C$, there must be $53 - 32 = 21$ who take $A$ and

124

$B$, but not $C$. Similarly, $55 - 32 = 23$ take $A$ and $C$, but not $B$, and $57 - 32 = 25$ take $B$ and $C$, but not $A$. So 21, 23, and 25 are placed in regions II, III, and IV, respectively.

Now the cases are considered for athletes taking only one vitamin. By adding the numbers already in $A$, we find $32 + 21 + 23 = 76$ of the 90 athletes are accounted for. Thus, $90 - 76 = 14$ athletes take $A$ alone. Similarly, $88 - 78 = 10$ take $B$ alone and $97 - 80 = 17$ take $C$ alone. So, 14, 10, and 17 are placed in regions V, VI, and VII, respectively.

We have now accounted for all the athletes who take vitamins. There are

$$14 + 21 + 10 + 23 + 32 + 25 + 17 = 142$$

of them, leaving $150 - 142 = 8$ athletes who take none of the three vitamins.

### Extension

This problem is worked exactly like the problem of the athletes and the vitamins. Students can first organize the information into a list:

    80% favor gun control
    53% favor nuclear power
    69% favor tighter pollution limits
    21% favor gun control and nuclear power
    46% favor gun control and pollution limits
    34% favor nuclear power and pollution limits
    9% favor all three

Now students draw a Venn diagram similar to the ones on the first student page and fill in the percentages from the information.

Because 9% favor all three, this percentage goes in region I. 21% favor gun control and nuclear power, and of these, 9% also favor pollution limits. Thus, $21\% - 9\% = 12\%$ favor gun control and nuclear power, but not pollution limits. (If students are confused by the percentages, have them think of there being 100 people in the survey. The percents then represent actual numbers of people—9% represents 9 people and so on.) Similarly, 37% favor gun control and pollution limits and 25% favor nuclear power and pollution limits. The Venn diagram is now filled in as shown:

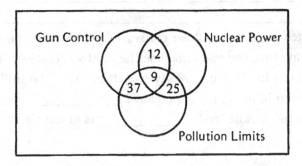

When students try to find the percentage of voters favoring only gun control, nuclear power, or pollution limits, they discover why Congressman Flipflop said the survey results were impossible. There are already 71% of the voters interviewed who favor pollution limits, but the survey said only 69% favored pollution limits. Thus, the results are not possible.

# I Always Lie

One of the most entertaining kinds of logic puzzles is the truth-teller/liar problem. A favorite one has an explorer discovering two tribes. The members of one tribe always tell the truth; those of the other tribe always lie.

Suppose you meet two tribesmen on a remote island and ask: "Are you a truth-teller?" The first man understands English, but he doesn't speak it. He answers, "Mahrd." The second tribesman interprets for you: "He says "yes," but he's a liar." Which man is the truth-teller and which is the liar? _____

To solve a problem such as this, ask yourself what the first man would answer if he were a truth-teller. _____ What would he answer if he were a liar? _____ Now show how the second man's answer tells you which man is in which tribe. _____

_____

_____

A somewhat more complicated puzzle uses three classes of people: one class always tells the truth, one always lies, and one sometimes lies and sometimes tells the truth. Suppose Lou always tells the truth, Marty sometimes tells the truth, and Pat always lies. Who is who in the following picture? _____ is at the left, _____ is in the middle, and _____ is at the right.

THE ONE IN THE
MIDDLE IS LOU

I'M MARTY

PAT IS IN
THE MIDDLE

*EXTENSION!* A state policeman drove into a small town and found it completely deserted except for five men in a jail cell. Some of the men in the cell were police officers and some were prisoners. A fight had broken out, and at the end of the fight the prisoners had thrown the keys and all the men's clothes and uniforms out the window. The state policeman knew that all the officers were truth-tellers and all the prisoners were liars. He asked only one question: "Which of you are policemen?"

A, the biggest man said, "I'm a policeman."

B said, "He'd say that anyway."

C, a man with an arm and a leg in casts, said, "Three of us are policemen."

D, the next man, said, "That's a lie."

E, the last man, said, "The big guy says he's policeman, but you'll have to decide whether or not he's telling the truth."

In about a minute the state policeman knew who were the policemen and who were the prisoners. How did he figure it out?

# Teacher's Notes for I Always Lie

*Truth-teller/liar problems are always a part of introductory logic courses. Part of the reason for this is that they lend themselves to so many amusing situations. A more important reason is that they offer such clear examples of problem solving via elimination. The examples given in this activity make a fine introduction for the more complex situation that is presented in "My Cat Rover."*

—————————————— NCTM Standards ——————————————

| 1 | 2 | 3 | 4 | 5 | 6 | 7 | 8 | 9 | 10 |
|---|---|---|---|---|---|---|---|---|---|
|   |   |   |   |   | • | • | • | • | • |

### Presenting the Activity

Discussing the title of this activity is a good way to open your presentation. "I always lie" is a nonsense statement—it can't be true. You may want to have your students suggest similarly impossible statements such as "I am dead" or "I have not yet been born." If either statement were true, the speaker could not, *by definition*, be making it.

Students realize, of course, that there are no tribes made up of perfectly consistent liars and truth-tellers, but this in no way diminishes their fun in considering the logical play that the situation affords.

The key to the first problem can be found as soon as one realizes that the response to, "Are you a truth-teller?" will always be "Yes." This is always the case–whether the respondent is in fact a truth-teller or a liar. We therefore have an effective yard-stick that we can use to measure the second man's answer. If he tells us that the first man answered "Yes," we know that he (the second man) is telling the truth. (This naturally brings us to the absurdity of the title. Can we imagine the first man answering that he's a liar?)

Some students will see that the logic involved here is analogous to multiplying a negative by a negative. Ask students to imagine themselves traveling in a country inhabited by truth-tellers and liars. The students know they are on a road that leads to the city they want to reach, but they don't know whether they're moving toward or away from that city. What question can they ask a native and be sure of getting the right answer—regardless of whether the native is a truth-teller or a liar? "What would you say if I asked you if this (pointing) were the right way to the city?"

Have students consider pointing (say) north, and asking. "Is this the right way to the city?" If that's the correct way, the truth-teller will answer "Yes," and if it's incorrect, he will answer "No." If you ask, "What would you say if I asked 'Is this the right way to the city?'," the truth–teller will answer the same way as before.

However, if you ask the liar, "Is this the right way to the city?" he will answer, "No," if you're pointing correctly. When you ask *"What would you say* if I asked, 'Is this the right way to the city?'," he must lie again. Thus, his answers are reversed to exactly those of the truth-teller. If the questioner is pointing in the right direction, both the truth-teller and the liar will say, "Yes." If he is pointing in the wrong direction, both will say, "No." In other words, the negation of a negative yields a positive.

The second problem is really only one step up the ladder as far as complication is concerned. Although this problem may seem too simple to require using a grid, it's good practice and good preparation for the Extension and problems such as "My Cat Rover." The steps, 1–4 as outlined to the next paragraph are shown on the grid:

|        | Lou | Pat | Marty |
|--------|-----|-----|-------|
| Left   | ×(2) | × | √(4) |
| Middle | ×(1) | √(3) | × |
| Right  | √ | × | × |

We see at the outset that the face in the middle can't be Lou's; if it were, he or she wouldn't say, "I'm Marty" (1). This tells us that the person on the left is lying (2), and because only one person is left, Lou must be the person on the right. Then, because we know Lou is a truth-teller, we take his word that Pat is in the middle (3). Again, only one is left and, therefore, Marty is on the left (4).

Using a grid allows students to keep track of things—usually unnecessary in puzzles this simple, but indispensible in more complicated problems.

### Extension

The Extension is a little more complex, but it uses the same reasoning students have used in the earlier problems. Thus, when A, the first man, says, "I'm a policeman," he is responding just as we've seen when the tribesman says he's a truth-teller. From this we see that both B and E are telling the truth, and, therefore, are policemen.

C and D introduce the complications. If C is telling the truth, then B, C and E are the policemen. However, we don't know this for sure; D says C is lying. If D is telling the truth, there must be four policemen. (We've already established that B and E are policemen, and if we add D, our total is three. Yet this can't be correct [in D's words], because C said three was correct.) For a total of four policemen we would then need A, B, D, and E.

We can't very well imagine C, with his broken arm and leg, overpowering the other four and tossing their keys and clothes out the window. Therefore, C is telling the truth: B, C, and E are the policemen, and A and D are the prisoners.

# My Cat Rover

I WONDER WHAT IT'S
LIKE TO LIVE IN A HOUSE
WITH **NO** KIDS

To be good at logic problems, you don't have to be good at algebra, but you do have to write things down. You have to keep track of things and ask one question at a time.

The following list comprises 16 sentences. Think of them as 16 pieces of information. From this information, you can figure out who owns Susan and who has no children.

1.  There are five houses in a row.
2.  The owners of the FORD live in the RED house.
3.  The VOLKSWAGEN owners have TWO children.
4.  The family that lives in the GREEN house has a pet named SPOT.
5.  The CHEVROLET owners' pet is called ROVER.
6.  The GREEN house is just to the right of the WHITE house.
7.  The family that owns a DUCK has FOUR children.
8.  A PIG lives in the YELLOW house.
9.  The family in the MIDDLE house has an animal named ROSIE.
10. The car at the FIRST house is a BUICK.
11. The CAT lives next door to the house with FIVE children.
12. The family that owns the PIG lives next door to the family with SEVEN children.
13. THOMAS, the GOAT, lives next door to SPOT.
14. The DODGE owners have a ZEBRA.
15. The BUICK's owners live next door to the BLUE house.
16. ROSIE's owners have a FORD.

Begin by setting up a table:

|  | House #1 | House #2 | House #3 | House #4 | House #5 |
|---|---|---|---|---|---|
| House Color |  |  |  |  |  |
| Car |  |  |  |  |  |
| Number of Children |  |  |  |  |  |
| Kind of Pet |  |  |  |  |  |
| Pet's Name |  |  |  |  |  |

Then put in what information you can be sure of. The rest will fall into place little by little. What does sentence 9 tell you? _____

What does sentence 10 tell you? _____

Using this and sentence 15, what color is house number 2? _____

Can house number 1 be white or green? _____ Why or why not? ___

_____

Can house number 1 be red? _____ Why or why not? _____

_____

Continue in this way until you have filled in the table. The two empty spaces will tell you where Susan lives and where no children live.

***EXTENSION!*** You can easily make up similar puzzles yourself. Start by making one that uses a 3 × 3 table instead of the 5 × 5 we've used here.

# Teacher's Notes for My Cat Rover

*People have enjoyed toying with logic problems for thousands of years. Although logic is an area of serious study, it's also one that lends itself to playfulness. No operational algebraic skills are needed for this problem, but both patience and analysis are required. For the best results, give "My Cat Rover" as a homework assignment. Then allow a full class period the following day to work through the problem.*

*(We're very grateful to Patrick Sneeringer of Nevada Union High School for updating this old war horse.)*

---

NCTM Standards

1    2    3    4    5    6    7    8    9    10
                          •                          •

---

### Presenting the Activity

To some, the problem will seem overwhelming at first. This is simply because of the amount of information to be juggled. It is for this reason that we suggest assigning the problem well ahead of discussing it. This allows some time for the students to familiarize themselves with the information.

The completed table is

|  | House #1 | House #2 | House #3 | House #4 | House #5 |
|---|---|---|---|---|---|
| House Color | Yellow | Blue | Red | White | Green |
| Car | Buick | Chevrolet | Ford | Volkswagen | Dodge |
| Number of Children | 5 | 7 | 4 | 2 |  |
| Kind of Pet | Pig | Cat | Duck | Goat | Zebra |
| Pet's Name |  | Rover | Rosie | Thomas | Spot |

Sentence 9 is a good place to start. It allows students to enter "Rosie" at the bottom of the House #3 column. Sentence 10 then places "Buick" in the first column. Then sentence 15 identifies house number 2 as blue.

Whereas the green and white houses are adjacent (sentence 6), neither of them can be house number 1. House number 1 must be either yellow or red. From sentences 2 and 10 we see that house number 1 must be yellow. Sentence 8 tells us to put the pig in the first column. From this and sentence 12, we see that seven children are placed in column 2.

Again looking at sentence 6, we can see that the red house must be either number 3 or number 5. Red can't be house number 4, because green and white are next to each other. What happens if we make house number 5 red? This can't be. Sentences 2 and 16 place the red house, Rosie, and the Ford in column 3. This then gives us all our house colors: from left to right, yellow, blue, red, white, and green.

Sentence 4 tells us that Spot goes in column 5, and sentence 13 tells us that Thomas and goat go in column 4. Now students should analyze the information in sentence 5. The Chevrolet and Rover must go in column 1 or column 2 because the pets are all named

132

in columns 3, 4, and 5. However, the car in the first column is a Buick, so Chevrolet and Rover are in column 2.

Three houses now have their cars identified. Thus, the Volkswagen and Dodge must go in columns 4 and 5. By sentence 14, the Dodge owners have a zebra. Thus, the Dodge cannot be in column 4 because the goat lives there. The Dodge and zebra are in column 5 and the Volkswagen is in column 4.

Sentence 3 now places two children in column 4, and sentence 11 shows that the cat has to be in column 2. The duck is the only animal left, so it goes in column 3. From sentence 7 we see that four children are placed in column 3 and, thus five children in column 1, next door to the cat.

In the completed table, the two empty spaces give us our answers: Susan, the pet pig, lives in the yellow house and there are no children at the green house.

### Extension

Some of your students will probably discover that the fastest method is simply to change the categories used in the given puzzle. They can use such things as hobbies, favorite foods, nationalities, felonies committed, and so on. These can then be directly plugged into the statements given on the student page.

Starting from scratch, a 3 × 3 table is, of course, much easier to construct than a 5 × 5. The trick is to fill in the table first and *then* make up the statements.

### Alternate Extension

Change the categories as suggested in the Extension and shuffle the order of the clues. In addition, delete sentence 16 and have sentence 13 read simply "The goat's name is Thomas" (or whatever category now fits). This will then require some trial-and-error testing from about the midpoint of the problem on.

# Algebraic Fallacies

At some time or another, almost every student of algebra comes across a "proof" that $1 = 2$ or $1 = 3$ or other such nonsense. One such proof goes like this:

1. Let $\quad\quad\quad\quad\quad\quad\quad\quad$ $a = b$
2. Multiply both sides by $a$. $\quad$ $a^2 = ab$
3. Subtract $b^2$ from both sides. $\quad$ $a^2 - b^2 = ab - b^2$
4. Factor. $\quad\quad\quad\quad\quad\quad\quad$ $(a + b)(a - b) = b(a - b)$
5. Divide both sides by $(a - b)$. $\quad$ $(a + b) = b$
6. Because $a = b$, then $\quad\quad$ $2b = b$
7. Divide both sides by $b$. $\quad\quad$ $2 = 1$

Does 1 really equal 2, or is there a fallacy somewhere in the preceding steps? If there is a fallacy, where does it occur and why is it fallacious or invalid?

---

A similar process can be used to prove that any two unequal numbers are equal:

1. Assume that $\quad\quad\quad\quad\quad\quad$ $x = y + z$
   If $x$, $y$, and $z$ are positive numbers,
   this implies $x > y$.
2. Multiply both sides by $x - y$. $\quad$ $x^2 - xy = xy + xz - y^2 - yz$
3. Subtract $xz$ from both sides. $\quad$ $x^2 - xy - xz = xy - y^2 - yz$
4. Factor. $\quad\quad\quad\quad\quad\quad\quad$ $x(x - y - z) = y(x - y - z)$
5. Divide both sides by $(x - y - z)$. $\quad$ $x = y$

134

Thus $x$, which was assumed to be greater than $y$, has been shown to equal $y$. Again, where and why does the fallacy occur?

---

A different kind of fallacy can be used to show that $4 = 8$. Consider these equations:

1. Given $\qquad\qquad\qquad\qquad\qquad\qquad$ $16 - 48 = 64 - 96$

2. Add 36 to both sides. $\qquad\qquad\qquad\quad$ $16 - 48 + 36 = 64 - 96 + 36$

3. Each member of the equation is now a perfect square, so that $\qquad$ $(4 - 6)^2 = (8 - 6)^2$

4. Take the square root of both sides. $\quad$ $4 - 6 = 8 - 6$

5. Therefore, $\qquad\qquad\qquad\qquad\qquad$ $4 = 8$

What is the fallacy this time?

---

***EXTENSION!*** Where and how does the fallacy occur in the following equations?

$$\sqrt{-1} = \sqrt{-1},$$
$$\sqrt{\frac{1}{-1}} = \sqrt{\frac{-1}{1}},$$
$$\frac{\sqrt{1}}{\sqrt{-1}} = \frac{\sqrt{-1}}{\sqrt{1}},$$
$$\sqrt{1} \cdot \sqrt{1} = \sqrt{-1} \cdot \sqrt{-1},$$
$$1 = -1.$$

135

# Teacher's Notes for Algebraic Fallacies

*Mathematics students often make errors in their work that are more basic than computational errors. When the theories behind mathematical operations are poorly understood, there is a good chance that operations will be applied illogically. Students unaware of certain limitations on these operations are likely to use them where they do not apply. The following paradoxes illustrate how fallacies arise in algebra when algebraic operations are applied beyond their limitations.*

—————————————— NCTM Standards ——————————————

| 1 | 2 | 3 | 4 | 5 | 6 | 7 | 8 | 9 | 10 |
|---|---|---|---|---|---|---|---|---|----|
|   | • |   |   |   |   |   |   |   |    |

### Presenting the activity

Ask students to analyze the first "proof" and find out where the reasoning breaks down. Of course, the trouble is in the fifth step. Because $a = b$, then $a - b = 0$. Therefore, division by zero was performed, and this is not permissible.

Discuss what division means in terms of multiplication. To divide $a$ by $b$ implies that there exist a number $y$ such that $b \cdot y = a$ or $y = \frac{a}{b}$. If $b = 0$, there are two possibilities: either $a \neq 0$ or $a = 0$. If $a \neq 0$, then $y = \frac{a}{0}$ or $0 \cdot y = a$. Ask your students if they can find a number that, when multiplied by zero, will equal $a$. Your students should conclude that there is no such number $y$.

In the second case, where $a = 0$, $y = \frac{0}{0}$ or $0 \cdot y = 0$. Here, any number for y will satisfy the equation. Hence, any number multiplied by zero is zero. Therefore, we have the "rule" that forbids division by zero.

There are other fallacies based on division by zero. In the second proof, the fallacy occurs in the division by $(x - y - z)$, which is equal to zero.

We can even "prove" that all positive whole numbers are equal. By doing long division, we have, for any value of $x$,

$$\frac{x - 1}{x - 1} = 1,$$

$$\frac{x^2 - 1}{x - 1} = x + 1,$$

$$\frac{x^3 - 1}{x - 1} = x^2 + x + 1,$$

$$\frac{x^4 - 1}{x - 1} = x^3 + x^2 + x + 1,$$

$$\vdots$$

$$\frac{x^n - 1}{x - 1} = x^{n-1} + \cdots + x^2 + x + 1.$$

Let $x = 1$ in all of these identities. The right side then assumes the values $1, 2, 3, 4, \ldots, n$. The left side members are all the same. Consequently, $1 = 2 = 3 = 4 = \cdots = n$. In this example, the left side of each of the identities assumes the value $\frac{0}{0}$ when $x = 1$. This problem shows that $\frac{0}{0}$ can be any number.

Another class of fallacies includes operations that ignore that a quantity has two square roots of equal absolute value, one positive and the other negative. In the third proof, students should discover that the fallacy lies in taking the improper square root. The correct answer should be $(4 - 6) = -(8 - 6)$.

The following fallacies are based upon failure to consider all roots of a given example.

Have students solve the equation $x + 2\sqrt{x} = 3$ in the usual manner. The solutions are $x = 1$ and $x = 9$. The first solution satisfies the equation, whereas the second solution does not. Have students explain where the difficulty lies.

A similar equation is $x - a = \sqrt{x^2 + a^2}$. By squaring both sides and simplifying, we get $-2ax = 0$ or $x = 0$. Substituting $x = 0$ in the original equation, we find that this value of $x$ does not satisfy the equation. The root chosen should have been $-(x - a)$.

### Extension

So far we have dealt with square roots of positive numbers. Ask students what happens when we apply our usual rules to radicals that contain negative numbers, in light of the extension problem. In this proof have students replace $i$ for $\sqrt{-1}$ and $-1$ for $i^2$ to see where the flaw occurs.

Another proof that can be used to show $-1 = 1$ is the following. Students have learned that the product of two square roots is the square root of the product. For example, $\sqrt{2} \cdot \sqrt{5} = \sqrt{2 \cdot 5} = \sqrt{10}$, but this then gives $\sqrt{-1} \cdot \sqrt{-1} = \sqrt{(-1)(-1)} = \sqrt{1} = 1$. However, $\sqrt{-1} \cdot \sqrt{-1} = (\sqrt{-1})^2 = -1$. It therefore may be concluded that $1 = -1$, because both equal $\sqrt{-1} \cdot \sqrt{-1}$. Students should try to explain the error. They should realize that we cannot apply the ordinary rules for multiplication of radicals to imaginary numbers.

# CHAPTER 7

# Linear Programming

- Graphing Linear Inequalities
- Making the Most of It
- Food for Thought

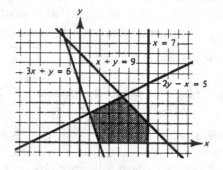

Linear programming did not really become a subject in its own right until the late 1950s. Kemeny's text (1956) is generally regarded as the watershed. The principles and techniques of linear programming had been around for centuries, of course; it just hadn't been practical to pursue them. Most real applications of linear programming require scores or hundreds of equations. Only the advent of the computer clarified the advantages of what, until then, had been an amusing side issue in mathematics.

The situations in these activities use only a few equations. However, the principles and criteria that are applied are exactly those used in complex commercial, industrial, and scientific situations. In addition, students are introduced to some of the terms special to linear programming.

"Graphing Linear Inequalities" is a prerequisite to linear programming. It allows students to define areas of feasible solutions. Many texts include a section on graphing linear inequalities, and of course your text's presentation can be an activity used in place of this. However, "Graphing Linear Inequalities" contains all the information necessary for the following two activities and it is recommended that students work through it before tackling "Making the Most of It" and "Food for Thought."

"Making the Most of It" presents a simple yet realistic problem in linear programming. It is set in a situation that high-school students can readily identify with and it's a clear, uncomplicated introduction to a powerful mathematical tool. "Making the Most of It" is another eye-opener for students.

"Food for Thought" is indeed that. It takes the same principles used in "Making the Most of It" and poses a problem that's in a more complicated setting and that requires students to work more independently. Students working through this problem will have to be careful. However, you can assure students who do successfully complete this activity that they will have no difficulty in standard, college-level linear programming courses.

## One Final Note

As the title of this series and several preceding statements indicate, these activities are meant to be motivational—to kindle the interest—largely through the variety and novelty of problems that your students have not seen before. An interesting facet of motivation is how very contagious it is. Therefore, as a first cut, it is strongly recommended that you simply browse through this book and make a mental note of the activities that are especially appealing to you yourself. Then make the decision as to where in your course these activities fit best. Your own enthusiasm will assure a successful and enjoyable learning experience for your class.

# Graphing Linear Inequalities

Many practical problems involve finding maximum profits or minimum costs. These kinds of problems are solved by using *linear programming*. To solve linear programming problems, you have to be able to graph linear inequalities.

To graph the inequality $4y - 3x \geqslant 12$, begin by graphing the equation $4y - 3x = 12$ on the following grid. This line is a *boundary* line and includes all the solutions of $4y - 3x = 12$.

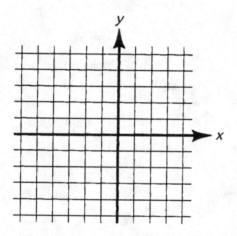

The graph of the inequality $4y - 3x \geqslant 12$ includes this boundary line and also the half-plane on one side of the line. To find out which half-plane is included, test a point not on the boundary line to see if it is a solution of the inequality. The origin, $(0, 0)$, is the easiest point to test.

Is $(0, 0)$ a solution of $4y - 3x \geqslant 12$? _____

Which half-plane should be included? _____

Shade this region of the graph.

Now graph the following *system* of linear inequalities:

$$x + y \leqslant 5,$$
$$2x - y \leqslant 4.$$

First graph the boundary lines for each inequality on the following grid.

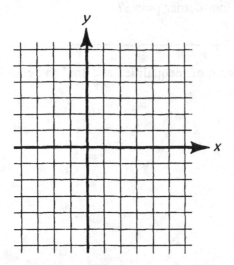

What are the equations of the boundary lines? _____

_____

Next, decide which half-plane should be included for each inequality. The graph of the system is the region where the two half-planes overlap. Shade this region.

How could you find the coordinates of the point where the boundary lines intersect? _____

_____

What is this point? _____

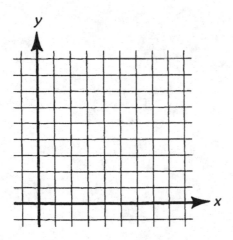

Now let's try a system of four linear inequalities. On the preceding grid, graph the system

$$5x + 2y \leqslant 30,$$
$$4y - x \leqslant 16,$$
$$x \geqslant 0,$$
$$y \geqslant 0.$$

141

This region is a quadrilateral whose four vertices may also be called *corner points*. What are the coordinates of the four corner points?

---

***EXTENSION!*** Graph the following system of inequalities and find the coordinates of the corner points:

$$3x + y \geqslant 6,$$
$$2y - x \leqslant 5,$$
$$x + y \leqslant 9,$$
$$y \geqslant 0,$$
$$x \leqslant 7.$$

# Teacher's Notes for Graphing Linear Inequalities

*Most textbook graphing exercises are simply that—graphing exercises and nothing more. Little is said about the problem-solving applicability of graphs. This activity introduces the idea of an infinite area of feasible solutions and, of course, is a necessary lead-in to the following two activities on linear programming. Students should be familiar with inequalities, and additional graph paper should be available to them.*

―――――――――――――― NCTM Standards ――――――――――――――

| 1 | 2 | 3 | 4 | 5 | 6 | 7 | 8 | 9 | 10 |
|---|---|---|---|---|---|---|---|---|----|
|   | • | • |   |   |   |   |   |   |    |

### *Presenting the Activity*

If students have been able to successfully graph linear equations, they should be able to graph linear inequalities easily. After students graph $4y - 3x = 12$, they test $(0, 0)$ in $4y - 3x \geqslant 12$. Because $(0, 0)$ is not a solution of the inequality, students shade the region above the line as shown:

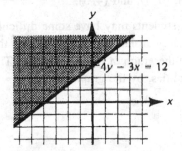

Explain that *all* the points on the line and in the shaded half-plane are solutions of the inequality.

Although students will be using only $\geqslant$ or $\leqslant$ activities in these linear programming activities, you may want to explain that in graphing $>$ or $<$ inequalities, the boundary line is drawn dashed. This indicates that the points on the line are *not* included in the graph; that is, these points are not solutions.

Students can now graph a system of linear inequalities. The equations of the boundary lines are $x + y = 5$ and $2x - y = 4$. After these lines are graphed, $(0, 0)$ can again be used as a test point. Because it is a solution of both inequalities, the graph is as shown:

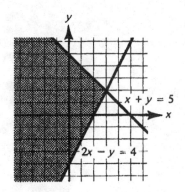

The point where the boundary lines intersect is found by solving the system of equations

$$x + y = 5,$$
$$2x - y = 4.$$

This point is $(3, 2)$. If the graph is drawn correctly, the point can be found directly from the graph.

To graph the system of four linear inequalities, students use the same method as shown already. The completed graph is

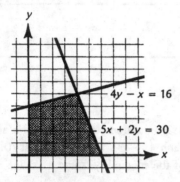

The coordinates of the corner points are $(0, 0)$, $(6, 0)$, $(0, 4)$, and $(4, 5)$.

When the system includes four or more inequalities, students may have some difficulty keeping track of which side of the various boundary lines is shaded. One way to do this is to use arrows to show which side of each line is in the graph of the solution set. This method is illustrated, using the system of four inequalities, by the graph.

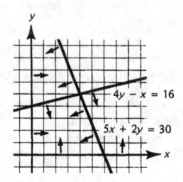

Another method would be to shade the half-plane that is *not* included in the graph of each inequality. Then, the *unshaded* region would show the solution of the system.

The graph of the system of five inequalities is

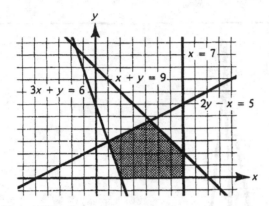

The coordinates of the corner points are $(2, 0)$, $(7, 0)$, $(7, 2)$, $(1, 3)$, and $(4\frac{1}{3}, 4\frac{2}{3})$. Notice that the coordinates of the last point cannot be read from the graph. They are found by solving $x + y = 9$ and $2y - x = 5$.

# Making the Most of It

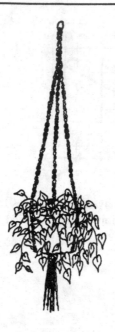

The 4-H Club wants to go to the state fair and needs to raise money for travel expenses. Nine members decide to make planter boxes and plant hangers to sell. A lumber yard and a crafts store donate enough wood and rope to make 30 boxes and 40 hangers. Each member can spend 20 hr making boxes and hangers. They discover a box takes 4 hr to make and a hanger takes 3 hr. They decide to charge $10 for a box and $8 for a hanger. They feel they will be able to sell everything they make. How many boxes and how many hangers should they make to maximize their profits?

Begin by translating the information in the problem to a system of inequalities. Let $x$ equal the number of planter boxes and let $y$ equal the number of hangers. They can make at most 30 boxes and 40 hangers, so

$x \leqslant$ _____ and $y \leqslant$ _____.

What is the total number of working hours? _____ Whereas a box takes 4 hr and a hanger takes 3 hr,

$4x + 3y \leqslant$ _____.

The number of boxes and the number of hangers cannot be negative, so

$x \geqslant$ _____ and $y \geqslant$ _____.

Now consider the profit, $P$. They charge $10 for a box and $8 for a hanger, so

$$P = \underline{\hspace{1.5cm}} x + \underline{\hspace{1.5cm}} y.$$

We want to find the maximum value for $P$ that is also a solution of the preceding system of five inequalities. To do this, first graph the five inequalities on the following grid:

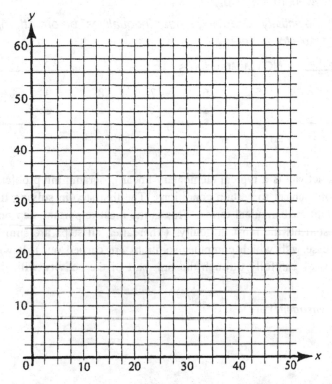

Every point in the shaded region of your graph is a solution of the system of inequalities. To find the one that gives a maximum value of $P$ seems like a difficult task, but is really pretty easy. It has been proved that *the maximum and minimum values occur at the corner points of the region.*

The coordinates of the corner points are _____.

Substitute each of these points in your profit equation. How many boxes and how many hangers should be made to maximize the profits? _____

***EXTENSION!*** The 4-H Club members decide to make a different planter box. This one will take 5 hr to make and there is enough wood for only 27 of them. However, the members will be able to charge $15 for each. Now how many boxes and hangers should they make to maximize profits?

# Teacher's Notes for Making the Most of It

*Although linear programming is a topic not usually treated until second-year algebra, it can be successfully presented any time after students are comfortable with systems of linear equations. Although the problems presented here are barely an introduction to the subject, most students are truly amazed at what complex problems simple linear inequalities allow them to solve.*

*Students should be comfortable with the activity "Graphing Linear Inequalities" before attempting this one. Additional graph paper should be available to students.*

―――――――――――― NCTM Standards ――――――――――――

| 1 | 2 | 3 | 4 | 5 | 6 | 7 | 8 | 9 | 10 |
|---|---|---|---|---|---|---|---|---|----|
|   | • | • |   |   | • |   | • | • | •  |

### Presenting the Activity

The problem posed in this activity is a typical elementary linear programming problem. Its aim is to maximize profit subject to certain conditions. The first step in solving the problem is to translate it into mathematical terms. This is the most important step and is also the one that gives students the most difficulty. (Thereafter, all steps are simply mechanical, and thus the ease with which computer solutions are carried out.) It will probably be necessary to work carefully through the first half of the activity with the class.

The five inequalities (the *constraints*) are

$$x \leqslant 30,$$
$$y \leqslant 40,$$
$$4x + 3y \leqslant 180,$$
$$x \geqslant 0,$$
$$y \geqslant 0.$$

The profit equation is $P = 10x + 8y$.

Now students graph the five inequalities as shown:

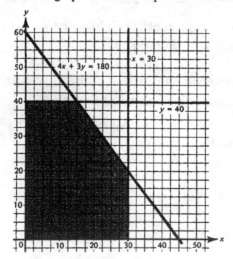

148

The shaded region (the *feasible region*) contains all the points that satisfy all five inequalities.

The coordinates of the five corner points can be read directly from the graph. They are: $(0, 0)$, $(30, 0)$, $(30, 20)$, $(15, 40)$, and $(0, 40)$. Because the maximum and minimum values must occur at the corner points, these values are substituted in the profit equation:

$(0, 0)$: $P = 10(0) + 8(0) = 0$,
$(30, 0)$: $P = 10(30) + 8(0) = 300$,
$(30, 20)$: $P = 10(30) + 8(20) = 460$,
$(15, 40)$: $P = 10(15) + 8(40) = 470$,
$(0, 40)$: $P = 10(0) + 8(40) = 320$.

Thus, the maximum profit is made when $x = 15$ and $y = 40$. The club members should make 15 boxes and 40 hangers.

### Extension

The Extension requires changing the profit equation and two of the inequalities. The five inequalities are now

$$x \leqslant 27,$$
$$y \leqslant 40,$$
$$5x + 3y \leqslant 180,$$
$$x \geqslant 0,$$
$$y \geqslant 0.$$

The profit equation is $P = 15x + 8y$. The graph of the inequalities is

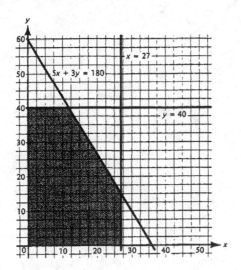

The corner points are: $(0, 0)$, $(27, 0)$, $(27, 15)$, $(12, 40)$, $(0, 40)$. Substituting these in the profit equation gives

$(0, 0)$: $P = 15(0) + 8(0) = 0$,
$(27, 0)$: $P = 15(27) + 8(0) = 405$,

149

$(27, 15)$: $P = 15(27) + 8(15) = 525,$

$(12, 40)$: $P = 15(12) + 8(40) = 500,$

$(0, 40)$: $P = 15(0) + 8(40) = 320.$

Thus, the maximum profit is obtained when the club members make 27 boxes and 15 hangers.

# Food for Thought

Nutritionists indicate that an adequate daily diet should provide at least 75 g of carbohydrate, 60 g of protein, and 60 g of fat. An ounce of food A contains 6 g of carbohydrate, 2 g of protein, and 3 g of fat. An ounce of food B contains 3 g of carbohydrate, 4 g of protein, and 3 g of fat. Also, the number of calories per day should not be more than 1500. An ounce of food A contains 75 calories and an ounce of food B contains 50 calories. Food A costs 10¢ an ounce and food B costs 15¢ an ounce. How many ounces of each food should be combined per day to meet the nutritional requirements at the *minimum* cost?

Translate the information to a system of inequalities. Let $x$ equal the number of ounces of food A per day and let $y$ equal the number of ounces of food B. What is the system of inequalities?

_____

_____

_____

Graph these inequalities on the accompanying graph.

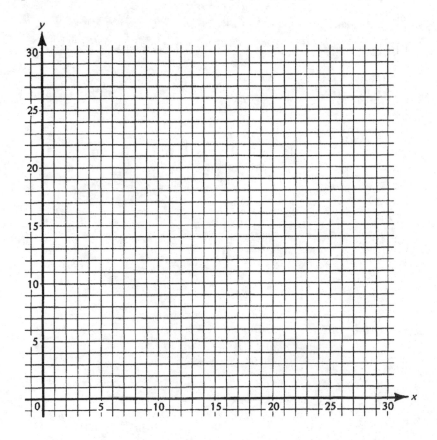

If $C$ is the total daily food cost,

$$C = \underline{\hspace{3cm}} .$$

What are the coordinates of the five corner points? $\underline{\hspace{5cm}}$

$\underline{\hspace{10cm}}$

Substitute each point in the cost equation. Which values of $x$ and $y$ *minimize* the cost? $\underline{\hspace{7cm}}$

Suppose you want to meet the nutritional requirements and spend exactly $2.70 per day. How many ounces of each food can be combined for $2.70 per day?

$\underline{\hspace{3cm}}$

***EXTENSION!*** Suppose a person can stand no more than 500 units per day of vitamin A and no more than 400 units per day of vitamin B. Also, the total number of units per day of the two vitamins must be at least 300 units and no more than 800 units. In addition, the number of units per day of vitamin B must be at least one-half but no more than three times the number of units per day of vitamin A. Suppose vitamin A costs 0.07¢ per unit and vitamin B costs 0.05¢ per unit. How many units of each vitamin should be taken per day to meet the given standards at the minimum cost?

# Teacher's Notes for Food for Thought

*This activity uses exactly the same concepts and operations required by "Making the Most of It." However, the problems here are more complex and students are expected to work more independently. Thus, "Food for Thought" is probably best assigned only to your superior students and, of course, those who have successfully completed "Making the Most of It." Additional graph paper should be available.*

_____ NCTM Standards _____

| 1 | 2 | 3 | 4 | 5 | 6 | 7 | 8 | 9 | 10 |
|---|---|---|---|---|---|---|---|---|----|
|   | • | • |   |   | • |   | • | • | •  |

### Presenting the Activity

An adequate daily diet must contain at least 75 g of carbohydrate and food A contains 6 g and food B contains 3 g. The inequality that expresses this information is $6x + 3y \geqslant 75$ or $2x + y \geqslant 25$. Similarly, for the protein, $2x + 4y \geqslant 60$ or $x + 2y \geqslant 30$; for the fat, $3x + 3y \geqslant 60$ or $x + y \geqslant 20$. The number of calories per day should not be more than 1500, so $75x + 50y \leqslant 1500$ or $3x + 2y \leqslant 60$. Also, the amounts of foods A and B cannot be negative, so $x \geqslant 0$ and $y \geqslant 0$. The system of inequalities and its graph are as shown:

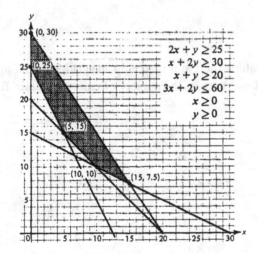

Each ounce of food A costs 10¢ and each ounce of food B costs 15¢, so $C = 10x + 15y$.

The coordinates of the five corner points are shown on the graph. The point $(15, 7.5)$ is found by solving $x + 2y = 30$ and $3x + 2y = 60$. The other points can be found directly from the graph. Substituting the coordinates in the cost equation, students find the minimum value of $C$ is $2.50 for $(10, 10)$. Thus, 10 ounces of food A and 10 ounces of food B should be combined per day.

To find the number of ounces of each food to be combined for $2.70, substitute $2.70 for $C$ in the cost equation and graph the line $270 = 10x + 15y$. (Remind students that because the coefficients of $x$ and $y$ are in cents, $2.70 must also be in cents.) This line is the heavy line shown in the next graph. Every point both on the line and also in the feasible region satisfies the conditions. There are three integral solutions: $(12, 10)$, $(9, 12)$, and $(6, 14)$.

153

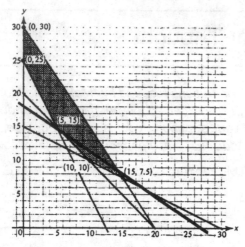

This graph can be used to introduce another way to find maximum or minimum values. If the cost $C$ is fixed at a particular value, we get a linear equation whose graph is called an *isoline*. The graph of $270 = 10x + 15y$ is an isoline. All points along this line have the same cost. Also, all isolines are parallel. As $C$ decreases, the isolines move closer to the origin, and as $C$ increases, the isolines move further from the origin. The last point in the feasible region that the isolines touch moving *closer* to the origin is the *minimum* value for the inequalities. The last point touched moving *further* from the origin is the *maximum* value. This method for finding maximum or minimum values is particularly useful if the feasible region has a large number of corner points.

### Extension

Students again begin by translating the information to a system of inequalities. Letting $x$ equal the number of units of vitamin A and $y$ equal the number of units of vitamin B, the system and its graph are

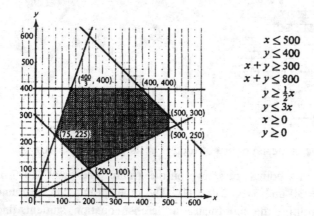

$$x \leq 500$$
$$y \leq 400$$
$$x + y \geq 300$$
$$x + y \leq 800$$
$$y \geq \tfrac{1}{2}x$$
$$y \leq 3x$$
$$x \geq 0$$
$$y \geq 0$$

The cost equation is $C = 0.07x + 0.05y$. The cost is minimized at $(75, 225)$, so 75 units of vitamin A and 225 units of vitamin B should be taken each day.

## CORWIN
## PRESS

**The Corwin Press logo**—a raven striding across an open book—represents the happy union of courage and learning. We are a professional-level publisher of books and journals for K-12 educators, and we are committed to creating and providing resources that embody these qualities. Corwin's motto is "Success for All Learners."